Handley
Regional
Library
In Memory of
Virginia "Jenny"
Smith
© Anke Blackburn

EARTHQUAKE STORMS

EARTHQUAKE STORMS

THE FASCINATING HISTORY AND VOLATILE FUTURE OF THE SAN ANDREAS FAULT

JOHN DVORAK

PEGASUS BOOKS
NEW YORK LONDON

EARTHQUAKE STORMS

Pegasus Books LLC
80 Broad Street, 5th Floor
New York, NY 10004

First Pegasus Books cloth edition February 2014

Interior design by Maria Fernandez

ISBN: 978-1-60598-495-7

10 9 8 7 6 5 4 3 2 1

Printed in the United States of America

Distributed by W. W. Norton & Company, Inc.

Dedicated to my father,
who always has a story to tell,
and to my mother,
who has always had the patience to listen to my father's stories.

CONTENTS

The whole world quaked . . .

—Homer, *The Iliad*

EARTHQUAKE STORMS

PROLOGUE
THE SWIMMER

Every earthquake is a surprise.
—Anonymous

I have often wondered—and envied—at the peculiar way Clarence Judson contributed to the study of earthquakes. A 36-year-old mechanic for United Railroads in San Francisco, he was following a daily routine of rising during the darkness of early morning and, after donning a robe, shoes, and a hat, careful not to wake his wife or two-year-old son or twin infant daughters, setting out for a brisk swim in the nearby waters of the frigid Pacific Ocean.

It was mid-April and the walk to the shore was a short one. The Judsons had lived for two years in a modest clapboard house two blocks from the ocean in what was a sparsely settled development known then and today as Ocean Beach.

On his way, Clarence Judson crossed a recently paved road that city planners had designated "The Great Highway"—a moniker that it still has today—then climbed over a short ridge of sand dunes and continued down to the beach.

He disrobed at the edge of the water and watched the waves before he entered. They were coming in sideways, in broken sets, the larger ones running up high, then pulling back. Judson later described them as "clawing at the beach."

He waded into the water up to his armpits and was ready to swim when a roller, larger than most, lifted him up, then set him down on his feet. A few moments passed and he prepared himself again to swim when, without warning and without any visible disturbance on the water surface, a great slap hit him across his entire body.

The slap stunned him and he started to go under. He struggled and somehow reached the surface just in time to suffer a second slap, then a third.

Now tumbling and his lungs filling with seawater, he fought desperately, able to summon enough energy to drag himself up onto the beach. Then the bizarre and most memorable part of his ordeal began.

Frightened by what had already happened, he tried to run to where his shoes and hat and robe lay, but discovered his legs refused to work. They shuddered uncontrollably—and he thought he must be paralyzed.

Surprised when he could move his legs again, he looked down and saw that the sand directly beneath his feet was filled with a phosphorous glow. Afraid his feet might be burned, he took off running, noticing that each step produced another incandescent strike.

When he reached his clothes, he tried to dress but was thrown down as the ground started to shake violently. This time he realized it was a terrible earthquake. His thoughts shifted to his family and he knew he had to get back to them.

The ground shook for almost a minute. As soon as it ended, Judson rose and dressed and raced back over the ridge of sand dunes and across the new paved road, which he noticed was badly cracked. He ran at top speed, past houses that were now tilted, some having shifted off their foundations.

When he arrived home, he found his wife frantic and his children crying. He tried to calm them, but a rumor soon came from a neighbor that said a giant sea wave might follow the quake. So the Judsons gathered what they could in blankets and, carrying these few possessions, joined other neighbors in an exodus, moving a few dozen blocks inland where they would camp for the next two nights.

As Clarence Judson and his family and their neighbors made their way, they could see dark clouds billowing up a few miles to the east. Later that afternoon, they would learn, as the rest of the world did, that San Francisco had been crippled that morning by a major earthquake—and that, because gas lines had been disrupted and chimneys had collapsed and the pipes that were to feed water to the city had broken, fires were now raging out of control, and the great city, home of nearly half a million, was burning to the ground.

Even today, much that Judson experienced that April morning in 1906—the wave slaps, the phosphorescent sand—has not been explained by science, an indication that our understanding of earthquakes is still in its infancy. And yet much has been learned since the infamous San Francisco earthquake, much of that knowledge gained by studying the San Andreas Fault, the same fault that ruptured that morning more than 100 years ago and caused roadways to crumble and houses to lean.

It was only after the 1906 San Francisco earthquake that most scientists finally realized that earthquakes were caused by the sliding of great crustal blocks against each other and not, as many had favored, by subterranean volcanic explosions. Four years after the San Francisco disaster, after a detailed study of the effects of the earthquake had been published, scientists also realized that earthquakes were powered by the release of elastic energy stored within the Earth, though how the energy had accumulated would remain a mystery for another half century. Not until the advent of the theory of plate tectonics in the 1960s would this and many other major mysteries about the Earth be solved.

Plate tectonics—the idea that the Earth's outer shell consists of a dozen or so mobile plates that collide or spread apart or slide horizontally past each other—would explain why earthquakes are so prevalent in California. In short, the tectonic plate that includes most of the continent of North America is moving, ever so slowly though constantly, westward, while the one that comprises most of the Pacific basin is moving, also at an excruciatingly slow pace, to the northwest. This difference in direction has caused a great fracture to form—the San Andreas Fault. It was movement along a segment of this fault that caused the 1906 earthquake.

Running for 800 miles from the redwood forests of Cape Mendocino southward to the rugged Sonoran Desert on the east edge of the Salton Sea near the border with Mexico, the San Andreas Fault passes beneath dozens of communities and close to two of the nation's largest cities, San Francisco and Los Angeles. It lies under major highways, pipelines, and crucial aqueducts. Scores of housing developments have been platted directly over it.

The fault is most readily apparent as an unusual alignment of river systems and valleys. In northern California, the Gualala and Garcia Rivers lie along it. In central California, the fault runs through narrow Bear Valley east of Point Reyes and through the fault's namesake, the San Andreas Valley. In the southern part of the state, picturesque Cuddy Valley and Leona Valley and arid Lone Pine Canyon mark its trend. The fault is responsible for the formation of both Cajon Pass and San Gregorio Pass, two vital corridors that link Los Angeles with the rest of North America. It is responsible for several desert oases and its trace can be seen, with a trained eye, from Palm Springs.

The words "San Andreas" are so well known that they have become synonymous with the stereotypical fast lifestyle that is

California—and with seismic destruction—though many perceptions about the fault are in error.

Technically, the San Andreas Fault is not *the* plate boundary between the North American and Pacific plates; instead, it is a major component of that boundary. The plate boundary actually reaches across a vast region that stretches from the Pacific coast to Colorado, New Mexico, and west Texas.

The San Andreas Fault is also a relatively young geological feature—as well as a transient one—when compared to other major geologic features, such as the Black Hills of South Dakota, which began to rise about 50 million years ago, or the basins filled by the Great Lakes, which began to sag hundreds of millions of years ago. By comparison, the current strand of the fault is a few million years old. The oldest segments of it are only 25 million years old and can be found in the mountains north and east of Los Angeles.

And, contrary to popular opinion, a major earthquake along the San Andreas Fault will *not* cause California to fall into the ocean. Instead, the San Andreas Fault and its many subsidiary faults are slowly tearing California apart, so that much of what is California today will be transformed into a collection of islands that are destined to be rafted northward across the Pacific.

Of obvious interest and concern is what will happen along the fault in the near future. On this question, scientists are in agreement: The last few *hundred* years have been a period of relative seismic calm in California. The calm cannot last, because the elastic energy that is building up as the Pacific and the North American plates grind against each other must be released. And that release can occur in only one way—as large earthquakes.

We also know from a study of the San Andreas Fault and its many subsidiary faults—the Hayward Fault, the Hollywood Fault, the Newport-Inglewood Fault, the San Jacinto Fault—that earthquakes do not occur randomly, nor do they reoccur like clockwork. Instead, large earthquakes can occur as clusters. And when a cluster of large earthquakes strikes over a period of, say, 100 years or so, there is an "earthquake storm."

Admittedly, such prolonged, intense periods of seismic activity in a single region are rare. A recent storm struck northern Turkey between 1939 and 1999 when 13 major earthquakes hit. One is probably going on now in central Asia where ten major earthquakes have hit the Szechuan region since 1893. In both cases—in Turkey along the North Anatolian Fault and in China along the Xianshuihe and Longmenchan Faults—the tectonic setting is similar to the San Andreas system of faults. And so an earthquake storm in California is a real possibility.

But how did we get to this realization? How did we arrive at our current understanding of the San Andreas Fault—and of earthquakes in general?

For the San Andreas Fault, it began decades before Clarence Judson made his fateful swim in the Pacific Ocean in 1906. In fact, it began, as much of the modern history of California does, with a quest for gold.

CHAPTER 1

A NOBLE EARTHQUAKE

We learn geology the morning after the earthquake.
—Ralph Waldo Emerson, 1860

On Wednesday morning, November 14, 1860, the *Golden Age*, a steamer from Panama, arrived in San Francisco. On the dock was a phalanx of city and state officials anxious to greet the ship's most anticipated passenger, Josiah Dwight Whitney of Northampton, Massachusetts.

Whitney, now 40, strong and stout with a ring of whiskers and a head of thinning hair, arrived with his wife of six years, the former Louisa Goddard of Manchester, England, their four-year-old daughter, Eleanor, and a long-time family maid, who, for the convenience of Mrs. Whitney's parrot, had agreed to be called "Mary."

Also traveling with Whitney were four young men who would be his assistants. There was Michael Eagan, who could do anything from camp cooking to laboratory chores. There were two recent college graduates, William Brewer and William Ashburner, who would serve as Whitney's scientific staff. The fourth man, only 19 and extremely near-sighted, was Chester Averill, whose family had sent him to California with Whitney as punishment for a student prank he had committed at Yale. Whitney would use him in a variety of ways—as a clerk, a mule driver, a barometrical reader, and a general factotum. Averill, for all his early misbehavior, would prove himself to be a most efficient and useful man.

At dockside, after the exchange of pleasantries and much fanfare, Whitney's wife, their daughter, and the maid were ushered to a private residence that had been prepared for them and where the Whitney family would reside for the next four years. Meanwhile, Josiah Whitney and his four assistants were taken to their new offices on the Montgomery block in the financial district of San Francisco so that they could begin work immediately—to compiling and completing a geological survey of the entire state of California.

The need for such a survey was self-evident among those who charged themselves with ensuring California's—and their own—financial future. The discovery of gold in 1848 had set off a race for riches, but gold production peaked quickly, so that by the end of the first decade it was barely half what it had been at the maximum. This worried the economic and political leaders of California who knew that new gold strikes were being made elsewhere—at the Comstock Lode in Nevada and near Pike's Peak in Colorado, both in 1858, and at Bodie, California, on the east side of the Sierra Nevada Mountains, in 1859—discoveries that convinced them that additional mineral riches must also still lie within their state. It was just a matter of finding them. But because it seemed that every sand bank of every stream had already been sluiced and every stone that lay in every outwash plain had been overturned and examined, it was agreed that a more concerted, less haphazard approach had to be taken. And so it was decided that the state of California would hire a bona fide rock expert. But how to find one?

Though several California miners clamored for the job, no less a political figure than the chief justice of the California Supreme Court, Stephen Johnson Field, decided he would search outside the state for the right man. Justice Field wrote to the presidents of several major colleges on the East Coast, asking who was the country's foremost authority on mineral ores. The responses were unanimous: Josiah Whitney, author of the widely acclaimed *The Metallic Wealth of the United States.*

Originally educated as a chemist at Yale College, after his graduation Whitney was taken aside by his father, who told his son that it was time he chose a profession, one that would insure sufficient income to support himself and a family. The law, his father said, would be an appropriate one. At first, Whitney dutifully followed his father's advice—until he met and had a private conversation with the man who was regarded as the greatest geologist of the age, Charles Lyell of King's College in London.

It was 1841 and Whitney was headed to Cambridge and preparing to enter Harvard Law School when he heard that Lyell, author of the influential and highly popular *Principles of Geology,* first published in 1830, was giving a lecture at the Odeon Theater in Boston. Whitney attended, as did more than a thousand other people. Lyell, unfortunately, had a cold that night and he spoke hesitantly and slowly. Yet despite these faults, Lyell displayed a clarity of thought that carried his audience, including Whitney. Afterwards, Whitney sought out the great man and spent an hour with him. Whitney left convinced that he should forgo a study of the law and pursue the more adventurous—though obviously less financially rewarding—science of geology. Perhaps, Lyell suggested, Whitney might make a living if he focused on the search for ore deposits.

Whitney took the advice and, though his father objected, spent the next five years in Europe, traveling through England, France, Germany, and Italy. He developed a special interest in mountains, crossing the Alps five times in five different places. He spent time in Russia, traveling as far east as Moscow. In all of these countries, he sought out authorities and discussed with them theories about the growth of mountains, the development of canyons, and the causes of eruptions and earthquakes. In 1847, six years after meeting Lyell, Whitney completed his tour of Europe and returned to the United States, where he found work searching for copper deposits on the upper peninsula of Michigan. That led him to compile all that was known of the mineral wealth of the United States, which he published in book form in 1854.

In 1855, he began working for the state of Iowa, looking for mineral wealth. In 1859, the state of Wisconsin hired him to search for iron and lead deposits. He spent winter months preparing reports at his home in Northampton, Massachusetts. In the spring of 1860, a letter arrived from Chief Justice Field offering him the newly created position of California's state geologist. Seeing an opportunity to work in the most mineral-rich state in the union, Whitney accepted immediately.

Just five days after his arrival in San Francisco, Whitney traveled by a special buggy to Sacramento for a private meeting with Governor John Downey, a Los Angeles man. During the meeting, the governor assured Whitney that he and his assistants could go anywhere—on public or private lands—request anyone's assistance, use any conveyance—they were given free passage on any train that ran within the state and on any ship that sailed along the coast or the inland waters—and could call upon any resource to complete their geologic survey and to prepare their reports.

The governor also had a personal request. Reports had recently arrived in Sacramento that a deposit of tin ore of "fabulous wealth" had been discovered in the southern part of the state near modern-day Riverside. Could Whitney keep the governor personally advised as to the potential of this discovery?

Whitney said he could not; he told the governor that whatever he learned about California's geology would be available to everyone equally and at the same time. His refusal—so William Brewer, one of the assistants, would write later—was the beginning of what, over the years, would develop into a full-blown antagonism between the governor and the men of California's Geological Survey.

But for now, as the geologic work was ready to commence, California politicians and businessmen and other prominent citizens came around often to the Montgomery office to become acquainted with Whitney and his assistants and to find out what they were doing—and to learn what they were discovering. Newspaper editors became especially fond of "the scientific men from Massachusetts," quoting them at every opportunity. By the end of their first month in California, the five men were already well known, though not one of them had yet sampled or studied a single rock.

By January 1861, Whitney and his assistants had assembled enough equipment to begin serious fieldwork. They had land-surveying instruments they'd brought from the East Coast that would be used to make topographic maps, a key element to conducting a geologic survey. In California, they purchased a medium-sized wagon with heavy braces, because few roads yet existed in the state. They also purchased a team of four strong mules to pull the wagon and an additional six mules for the men to ride and to carry extra equipment. Each man was issued a revolver and a large knife. Two carbines and two double-barreled rifles were included, considered essential parts of the camp gear. And each man was instructed that, whenever he left camp or went into a town, he was to carry a gun, knowing that if unscrupulous men saw that he was well armed, they would leave him alone.

Once they got under way, the five men soon discarded their East Coast attire, which had devolved into rags, torn apart by chaparral and jagged rocks, and adopted the rugged denim pants and rough flannel shirts of the gold miners. They also acquired cowhide boots and broad-brimmed hats. And though when he was at home Whitney insisted that clean sheets be placed on his bed every day and refused to wipe his hands twice on the same towel, when in the field he, as well as the others, learned to sleep in the open on oilcloth and to cover himself with a single blanket.

They cooked for themselves and ate whatever was at hand. They became skilled at washing clothes in muddy creeks and in carrying delicate mapping instruments up steep slopes in all weather conditions. In reading through their journals, one learns that they also came to handle, even to admire, the cantankerous California mule.

For four years, Whitney and his team of assistants crossed the state, descending into every major river valley and climbing to the crest of every major mountain range. Their travels and accomplishments were glorified in local newspapers. Whenever they passed close to a sizable town, community leaders sought them out, anxious to associate themselves with Whitney and his team of geologists.

In 1864, in recognition of his leadership, his men named the highest peak in the Sierra Nevada Mountains—which is also the highest point in the contiguous 48 states—Mount Whitney. The same year, this expert on mineral ores issued his first major report. Its contents shocked the people of California when they read it because, instead of telling where additional mineral deposits might be found, this 236-page tome described something else—*paleontology*!

The public's outcry was immediate: Why had so much public money been used to uncover the bones of long-extinct reptiles, to illustrate the imprinted skeletons of long-dead fish, and to describe the shells of useless clams? In scanning through the hundreds of pages of text based on four years of intense fieldwork, one finds the word "gold" only three times and the words "wealth" and "riches" not at all.

One outraged citizen carried the tome to the state assembly and antagonized legislators in their offices by reading sections to them. And the legislators reacted. They reduced both Whitney's annual salary and his budget by half. He responded by leaving the state and returning to Massachusetts, where Harvard College honored him with a professorship. It was there, in the halls of the nation's oldest and most distinguished academic institution, that Whitney prepared a second report that he entitled simply *Geology*.

The people of California read the second tome with great interest. At least this time there were nearly 200 pages—the middle third of the report—that described the gold-bearing regions of the state, focusing, not unexpectedly, on the Mother Lode, the 100-mile-long stretch at the base of the western Sierra Nevada Mountains where the most productive veins of gold had been discovered in January 1848. The report told how Whitney and his men had visited almost every major gold field in the state and how they had entered almost every major mine. The report described the geologic setting of each one, though here Whitney committed an unforgivable offense: Showing either an ignorance of or a total disregard for political realities, he pronounced more than half of the mining claims in the state as either worthless or unproductive. And, more than that, there was an important element missing from the report: There was no indication where additional gold might be found.

This lack of telling where new gold strikes might be made caused the state legislators to act again. This time, they cut Whitney's salary and his budget to zero. Now a Harvard professor, Whitney continued to live in Massachusetts, making an occasional trip to California to continue his fieldwork, his efforts now supported by private sponsors in New England and by benefactors of Harvard's museums.

In 1874, still without any new prospects as to where additional mineral ores might be found, the California state assembly voted to formally end the job of state geologist.

———

Today Whitney's *Geology* is regarded as a masterpiece, exquisitely written and describing not only the geology of California but also its fauna and flora, much of it now gone. Whitney recounts a visit to Yosemite when the valley was still pristine. He is the first to use the term "High Sierra." And he tells of personal adventures in confronting flash floods and assisting local lawmen who were searching for desperadoes. One of the few errors in his *Geology* is his assessment of the potential for an oil industry in California in which he stated that the oil-rich deposits known along the coast would never be of commercial value.

Because the survey work was begun during winter months, he and his assistants had started in the southern part of the state, where they took time, as the governor had requested, to investigate rumors of the discovery of a major tin deposit south of the San Bernardino Mountains.

By the time they arrived, hundreds of claims had already been made, covering all the hills and ridges for miles around. But as far as Whitney could determine, a tin-bearing mineral, cassiterite, could only be found at one spot: at the Cajalco Mine. And that mine, at the time he visited, consisted of a single shaft dug down only 36 feet and was mostly filled with water.

Nevertheless, Whitney collected rock samples from the Cajalco Mine and sent them to Boston and to New York to be assayed. The results were as he predicted: Though one sample yielded 60% metal, the others showed only small amounts. In all, Whitney thought the discovery was interesting—the Cajalco Mine is the only known occurrence of tin ore along the Pacific coast north of Mexico—but of little commercial value, a judgment borne out by a century of repeated attempts to extract ore from the mine.

It was fortunate—or perhaps unfortunate—that the first glimpse that Whitney had of California's geology was in the region around the Cajalco Mine because here the geology is extremely complex. There are no simple horizontal strata, arranged in layer-cake fashion, as visible in places like the Grand Canyon, indicative of what much of the earth looks like beneath our feet, and no simple upturned rocks and a few great folds that are so prevalent in the eastern half of the United States and across Europe, areas that Whitney had already studied. Instead, what he saw south of the San Bernardino Mountains was, in his words, "a Gordian knot," "a curious snarl." Today we understand the cause for this geologic maelstrom: It is the San Andreas Fault.

Whitney crossed the San Andreas Fault many times during his years of exploration, though he never recognized any of the features as indicating a geologic fault—nor would anyone for another 30 years. Today we see the same landscape and the same features with different eyes.

At San Bernardino, north of the Cajalco Mine, one can see three long mountain ranges. To the west are the San Gabriel Mountains. Immediately to the north and extending to the east are the San Bernardino Mountains. Separating them is a low area known as Cajon Pass, a corridor where vital lifelines pass—oil and water pipelines, electrical power lines, and the I-15 freeway that links Los Angeles and Las Vegas. Cajon Pass exists because the San Andreas Fault runs through it, forming the boundary between the San Gabriel and San Bernardino Mountains.

South of the city of San Bernardino are the San Jacinto Mountains, and they are separated from the San Bernardino Mountains by another low area, San Gregorio Pass, also a corridor used by pipelines and power lines and a freeway—the I-10 that connects Los Angeles and Palm Springs. Again, the San Andreas Fault is responsible for this pass and for the separation between the two mountain ranges.

But none of this was obvious to Whitney. As California's first geologist, he was perceptive enough to realize that there was something unusual here, something interesting. He also realized there were other geologic oddities in the state.

After completing a reconnaissance of the geology around the Cajalco Mine and San Bernardino, Whitney and his team of assistants moved west, following the southern base of the San Gabriel Mountains. They took time to pass through Los Angeles, then merely a settlement of a few thousand people, and inspect the sedimentary beds exposed by the Los Angeles River. From there, they continued to the Pacific Ocean and to Santa Barbara, where they picked up the old Spanish trail, *El Camino Real*, the King's Road, built a century earlier to connect Spanish missions, and headed north.

At Monterey, Whitney noticed that the old town was nestled against a block of white granite. He followed the granite south as far as the picturesque bay at Carmel, which he foresaw would "at some future time become a place of resort, if not for the fashionable world, at least for those who would combine, with the enjoyment of the ocean breezes, the pursuit of geological, botanical, or zoological studies."*

He tried to follow the granite farther south along the coast, but the cliffs were too steep, and he and his four men turned and continued north.

Again, along the coast just before reaching San Francisco, Whitney found another huge block of granite that, by all appearances, was identical to what he had seen at Monterey and Carmel. Later, working north of the Golden Gate, he found yet again huge blocks of the same granite, this time forming the peninsula of Point Reyes and the sea head at Bodega Bay and, 30 miles from the coast, comprising the Farallon Islands. Such blocks seemed to be scattered for hundreds of miles along the California coast.

Two summers later in 1862, he uncovered an even greater surprise. Whitney made his first trip to the gold mines of the Mother Lode. After that, he swung around the southern end of the Sierra Mountains and there, exposed in the core of California's highest mountain range, was the same white granite—identical in mineral content with what he had seen hundreds of miles away along the coast.

He took samples and sent them to Yale College for analysis, as he had done with samples of granite collected along the coast. The results were remarkable: By every measure the granite in the core of the Sierra Nevada Mountains is identical to that found along the California coast.

---

* Intriguingly, the white granite can be seen exposed at several famous golf courses in Carmel, including Pebble Beach and Cypress Point, the layouts of the fairways controlled by the contour of the granite.

How was this possible? He had no explanation.

Today, after a century and a half of additional work, which includes more refined chemical and mineralogical analyses of thousands of rock samples, extensive radiometric age dating of many of the same samples, and detailed geologic mapping that, in part, reveals a host of geologic structures, we know why: The blocks of granite that lie along the coast, which are part of a large feature known as the Salinian Block (named after the Salinas Valley north of Monterey), have actually *slid* hundreds of miles from where they formed at the southern end of the Sierra Nevada Mountains. Moreover, the sliding took place by a seemingly countless number of earthquakes that have occurred over eons along the San Andreas and nearby faults.

In all, this movement is a remarkable testament to the power and persistence of Earth's internal forces. A mountain range may get pushed up a few miles, but in California, whether one stands on a rocky outcrop of white granite at the edge of Monterey Bay or midway up the steep cliffs at Point Reyes, beneath one's feet is evidence for *hundreds* of miles of *horizontal* displacement. It is equivalent to a piece of real estate the size of Manhattan Island sliding from Pittsburgh to New York City.

But nearly a century would pass after Whitney's original work before this was clearly realized. During his era, what earthquakes were and what their accumulative effects could be was not understood. In fact, he and his contemporaries downplayed the role of earthquakes in geologic history. He himself did not even consider California to be especially prone to such events. In one of his annual addresses to the state assembly, he told legislators: "Luckily for us, California does not seem to be in the region of heavy shocks, so that whatever feeling of insecurity may have once existed on this account seems to have nearly died out."

Though he did concede, in his book *Geology*, that earthquakes, acting locally and in the distant past, were responsible for some of California's spectacular features such as Yosemite Valley, where strong seismic shaking had caused the valley floor to collapse. By and large, those in academic circles agreed with him.

But such ideas were woefully inadequate. And the fact that they were inadequate was pointed out by an unlikely source: a man of limited education who came to Yosemite Valley a few years after *Geology* was published. He had read the book thoroughly and proclaimed that Whitney's ideas about earthquakes and their consequences were wrong.

Born in Scotland in 1838, John Muir immigrated to the United States with his family at the age of ten. They settled on a small farm in Wisconsin where life was harsh, made all the more difficult for the boy due to the fact that his stern father punished him whenever he committed any perceived sin—and even when he had not.

In the 1860s, Muir attended classes for two years at the University of Wisconsin in Madison, receiving his only formal instruction in science. Here the professor of natural history was Ezra Slocum Carr, who often led his students on hikes through nearby fields and woods, investigating, as he put it, "Nature's basement rooms."

Of particular interest to Carr—and catching the attention of young Muir—was the abundance of geologic evidence for glaciers. There were rocky moraines that marked the terminal points of glacial lobes, deep grooves that recorded the scraping of rocks that had been pulled by slow-moving glaciers over a hard ground surface. There were also plenty of boulders—known as "erratics"—that had been transported great distances by flowing glaciers, then dropped as the surrounding ice melted away.

Muir, too poor to continue a formal education, found work at a factory that made parts for horse carriages. One day, as a result of his own admitted carelessness, he had a terrible accident. He was replacing a machine belt when, while holding a sharp file, the belt slipped, causing his hand to snap back and push the file into his right eye. Months of blindness followed. When his eyesight did recover, he vowed to forgo a life of pursuing comfort and security and instead see the beautiful parts of the world.

After a year of traveling, first by foot from Wisconsin to Florida then by ship to Panama and California, he found himself in San Francisco waiting for a ship to the Hawaiian Islands, when he decided to postpone that trip and remain in California a little longer. During his convalescence, a friend had brought him a stack of books and papers to read. Among them was a pamphlet complete with photographs that showed a beautiful place called Yosemite Valley. He decided he needed to see it before heading out into the Pacific.

He crossed San Francisco Bay on the Oakland ferry, then, armed with only a pocket map, he set out on foot for Yosemite, 200 miles away. He arrived at the entrance to the valley on July 11, 1868. He noticed there were boulders in a broad field, not piled up like debris but strewn around as if they had been dropped. Inside the valley, he recognized grooves scraped across bedrock, as he described them, "striated in a rigidly parallel way." He also found mounds that could only be glacial moraines. The grooves and the mounds of the boulders—erratics, similar to the ones in Wisconsin—were clear evidence that Yosemite Valley had once been "overswept by a glacier." Perhaps, Muir speculated, a huge glacier carved its way through during the Ice Age and dug out the valley itself. He decided to consult the only recognized authority: the first guidebook of Yosemite Valley, written by Josiah Whitney and much of it taken from his *Geology*. Muir was surprised by what Whitney had written.

Whitney acknowledged that glaciers had once capped the Sierra Nevada Mountains but, he reasoned, they could not have formed Yosemite Valley like Muir had speculated. His main argument against large-scale glacial action was this: Where is the material that glaciers supposedly removed from the valley? The erratics were just one sign of a possible glacier but there were no others, including a lack of huge moraines on the plain near the valley entrance.

Furthermore, Whitney discounted the idea that Yosemite Valley was a fissure in the Earth's crust: The valley was much too broad and flat-floored to support such an explanation. And it could not have formed by water erosion because mountain streams, like the Merced River, which flows through Yosemite, form narrow valleys, not broad ones with steep sides and flat floors. And besides, where was the material that river erosion would have removed?

So Whitney was left with only one geologic agent that could explain the existence of Yosemite Valley—an earthquake.

In Whitney's view, a single local and rare cataclysm—a powerful earthquake—had caused the floor of Yosemite Valley to drop suddenly, forming the spectacular vertical walls of Half Dome and El Capitan. It was such a cataclysm that explained why Yosemite Valley was unique in the Sierra Nevada Mountains.

Muir was unconvinced and responded to Whitney in his first published work, "Yosemite Glaciers," an essay that appeared on December 5, 1871, in the *New York Tribune*, the city's most widely read newspaper. He laid out the argument that "the Great Valley itself, together with all its domes and walls, was brought forth and fashioned by a grand combination of glaciers."

One wonders what direction the earthquake-versus-glacier debate might have taken if a major earthquake had not occurred beneath the Sierra Nevada Mountains less than four months later.

Muir was asleep in a small cabin he had built near Sentinel Rock on the valley floor when, during the early morning hours of March 26, 1872, he was awakened by shaking. "Though I had never before enjoyed a storm of this sort," he would later write, "the strange, wild thrilling motion and rumbling could not be mistaken."

He ran outside into a clear, moonlit night, feeling "both glad and frightened." "A noble earthquake," he shouted. He was sure he was going to learn something.

The shaking was so violent and varied, one pulse quickly succeeding another, that he had difficulty walking and had to balance himself as if on a ship in big waves. He could hear rocks from the steep walls descending all around him, and so he took shelter as best he could behind a large yellow pine. The shaking lasted almost two minutes. Muir watched as the face of Eagle Rock gave way and fell into the valley with a tremendous roar. But even before the great boulders had settled in place, he was on top of them, listening as they gradually quieted.

After sunrise, he walked about to see what other changes had occurred. He saw some people who were wintering in the valley assembled in front of a small hotel. He joined them and listened as they compared notes. They talked about the earthquake and whether they should leave the valley. A muffled rumbling similar to thunder was heard, followed by a shock milder than the initial one.

Muir recognized one of the people to be a shopkeeper in the valley who Muir knew was a firm believer in the cataclysmic origin of Yosemite Valley. Muir jokingly remarked that Whitney's "wild tumbled-down-and-engulfment hypothesis" might soon be given a test. Just then came another shock, and the shopkeeper became solemn. Muir now tried to cheer him up by telling him that the earthquake had just been Mother Earth "trotting us on her knee to amuse us and make us good." The man was not amused. He and the others quickly gathered their things and left the valley. A month later the man returned, but when another tremor happened the same day in a shocking coincidence, Muir watched as the shopkeeper was sent flying again.

Muir remained at Yosemite after the large earthquake and its aftershocks, intrigued by what might happen next. He filled a bucket with water and placed it on a table and watched it for hours, noticing that the surface of the water would dimple every time Mother Earth sent another slight tap from far below.

Whitney was in San Francisco during the earthquake and slept through the event, as did almost everyone else in the city. Thomas

Tennant, a self-appointed weather observer who had been in California since the Gold Rush, was one of the few in the Bay Area who felt the shaking, recording it as "a light earthquake."

By sunrise, telegraph wires were buzzing with reports. The earthquake had been felt from Shasta to San Diego and as far east as Eureka, Nevada. A quick study of the reports showed that the most severe shaking had not been in Yosemite Valley, but along the east side of the Sierra Nevada Mountains in Owens Valley at a small mining community known as Lone Pine.

By coincidence, Whitney was planning to work later that spring in Owens Valley, and so he passed through Lone Pine on May 21. He confirmed what had been reported in the newspapers: Almost every one of the 59 buildings of Lone Pine, mostly adobe houses, had collapsed. And 27 people, nearly 10% of the population, had been killed.

In an article he would write for *Overland Monthly*, a California-based magazine, he described entering Lone Pine as being "in the midst of a scene of ruin and disaster." He surveyed the damage, which was still quite evident. At some point he went to the north edge of town, where, even today, there is a poignant reminder of the sorrow an earthquake can bring. At the top of a sandy hill just west of State Highway 395 is a mass grave where 16 of the earthquake victims were buried. Today it is surrounded by a weathered wooden picket fence. The plot is so small that some of the bodies must have been laid atop others. At first thought, one is puzzled why so small an area was used as a grave site; then one remembers that all 16 had died at the same instant, and this was the only way, in this treeless expanse, to bury them quickly, something that was of prime importance in an era when disease spread rampantly.

From the grave site, one can see something else of curiosity related to the earthquake. To the southwest, in the direction of the Sierra Nevada Mountains, at the base of a range of low hills that, because of their striking appearance, has been the filming location for many popular Hollywood movies (including *How the West*

Was Won, Star Trek Generations, and *Transformers: Revenge of the Fallen,* as well as the opening scene of *Iron Man*) is a line that looks like a frozen wave of upturned rocks. Close inspection shows that is exactly what it is.

It is what geologists call a scarp, a step on the ground surface where one side of a fault has moved vertically with respect to the other, and it is as high as 20 feet, running for more than 40 miles parallel with the valley and the mountains. Along its face are huge boulders. People in Lone Pine told Whitney that the scarp formed during the earthquake, and he agreed.

The scarp was not the only feature that formed during the earthquake. There were also long cracks and fissures that Whitney found elsewhere on the valley floor. Following convention, he explained these and the 40-mile-long scarp as secondary effects of the earthquake—that is, they were the result of the violent shaking that had caused the loose ground to settle more in some places than others. If he had paused, he might have realized that there was a fundamental difference between the cracks and fissures and the rock-faced scarp—one that another man would clearly see and, thereby, change the whole perception of the cause and nature of earthquakes—but Whitney was not attuned to make such a distinction.

If he had been, he probably would have realized that a cataclysmic earthquake could not have formed Yosemite Valley, although after what had happened in 1872, he could be forgiven for thinking so.* But for him to have seen a distinction between what created the valley and an actual earthquake would have required

* In a similar vein, if Muir and others had known more about glaciers and their ability to scour and erode, he probably would not have proposed a glacial origin for Yosemite Valley. It would not be until 1913, when Francois Matthes, a geologist working in national parks, disposed of the earthquake theory and downplayed a glacial origin. Today, the general opinion is that Yosemite Valley is an erosional feature formed by river erosion and exfoliation of granite. The flat valley floor owes its existence to sediment trapped in shallow lakes that formed during a retreat of the glaciers.

a radical change in thinking. And to achieve that change, one had to know exactly what caused earthquakes, something Whitney himself still did not know. At the time, conventional wisdom held that such sudden and, at times, strong shakings were caused by volcanic explosions occurring inside the Earth.

In 1750, England, seldom disturbed by earthquakes, was shaken twice. The first occurred just after noon on February 8 and caused so much alarm that the barristers at Westminster thought the stone hall was falling down. Exactly a month later, on March 8 at half past five in the morning, another tremor struck, throwing people out of their beds and causing the chime hammers to strike bells in church steeples. The next day, John Wesley, renowned minister and founder of Methodism, gave his *Sermon 129, The Causes and Cure of Earthquakes.* In that sermon, he proclaimed, "God is himself the Author, and sin is the moral cause." The only cure, of course, was repentance.

Five years later, on November 1, 1755, an earthquake devastated Lisbon, Portugal. Tens of thousands died, most by drowning due to an earthquake-produced tsunami. Shaking was felt from Finland to North Africa and as far west as the Azores Islands. Then, 17 days later, just as news of the Lisbon disaster reached Boston, New England was struck by the strongest earthquake yet to hit that region.

The New England earthquake was felt from Nova Scotia to South Carolina. The most intense shaking was around Boston, where, though no one died, many brick walls and chimneys fell. John Adams, then 20 years old, was at his father's house in Braintree. The next day, presumably inspired by the earthquake, he started his famous diary, which would continue for more than 50 volumes and include details about the revolution and the difficulties of forming a constitutional government. The first entry, however, is about the earthquake, which caused his father's house "to rock and reel and crack as if it would fall in ruin around us."

John Winthrop, a professor of mathematics and philosophy at Harvard College, concerned that local clergy would replay Wesley's sermons, on November 26 gave "A Lecture on Earthquakes: Read in the Chapel of Harvard College." His lecture would mark a turning point in the study of earthquakes, suggesting that earthquakes could be explained based on physical causes, and denied that they were an intervention of God in earthly affairs.

During the lecture, he described the shock not as chaotic motion but as a "kind of undulatory motion," a kind of "wave of earth" like a wave of water—Winthrop being one of the first to provide such descriptions. As to a cause, he proposed that it was the result of a release of pressure created by a buildup of "fumes from fermenting minerals." More specifically, it was an explosion of gases produced by the underground mixing of iron and sulfur with water, a combination that chemists had already shown could produce an excess of heat—and, hence, might power volcanoes—and, if the explosion was confined, could shake the Earth.

More than a century earlier, French chemist Nicolas Lémery had given a demonstration of how the process might work. He mixed equal parts of iron shavings and powdered sulfur into a large jar filled with water until he had a paste. He then secured a lid on the jar, had assistants bury it a few feet underground, covered the jar with dirt, and waited. After a few hours, cracks formed on the ground surface over the buried jar. Gases emitted from the cracks could be ignited with glowing embers. There were even individual underground bursts that shook the ground.

This idea—that earthquakes were underground explosions caused by the mixing of naturally occurring chemical compounds—was the prevailing one when Whitney attended Yale College in the 1840s. By the 1870s, the idea had been tempered and some authorities were now suggesting that the sudden strong quaking of the ground surface might be a product of the slow, constant cooling of an initially hot Earth, which caused the entire planet to contract, shortening the surface much like an apple eventually shrivels and forms wrinkles on its skin. In truth, in the 1870s, no one knew what caused an earthquake except in a very general sense, as Whitney and others wrote that it was an "impulse" occurring somewhere inside the Earth.*

Having surveyed the damage at Lone Pine, seen the scarp, cracks, and fractures, and having rummaged through hundreds of telegraph wires and newspaper reports, Whitney concluded that the "impulse" that had produced the Owens Valley earthquake had originated beneath the axis of the Sierra Nevada Mountains, not realizing that the evidence of what had actually triggered the shaking—and the physical cause of all earthquakes—was actually in plain view to him, Muir, and countless others. But it required someone else with a different perspective to see the obvious, to understand the significance of the 40-mile-long scarp that wound its way close to the cemetery where the earthquake victims were buried.

As was said during one of the many laudatory remarks made immediately after his death, geologist Grove Karl Gilbert from Rochester, New York, had merely looked out the window of a train on which he was a passenger and from that came up with a new theory of mountain building.

It was 1871, a year before the Owens Valley earthquake, and he was on his first trip to the American West, riding on the recently

* Here there is a coincidence that spurred on the idea—at least to Whitney—that earthquakes were underground explosions. In July 1872, while Whitney was writing his report about the Owens Valley earthquake, a nitroglycerin factory exploded in San Francisco. Windows rattled and walls shook for miles around, prompting Whitney, who was in the city at the time, to write ". . . we have only to imagine an impulse given, like that produced by the nitroglycerin explosion—only on a vastly greater scale—to produce all the effects of the most disastrous shock."

completed transcontinental railroad. The train was passing through northern Nevada—on the route followed by I-80 today—and through the window he watched as the train made its way over a series of mountain ranges separated by basins. Conventional wisdom held that such a sequence of ranges and basins was the result of compression, such as in the Appalachians, where the Earth's crust had buckled and folded. But Gilbert saw no great folds in the mountains of Nevada. Furthermore, the mountain ranges gave the appearance of having been tilted, so that one side of a range was noticeably steeper than the other. To Gilbert, the explanation was obvious: A large area of the continent had been uplifted and then stretched.

The alternating pattern of mountain range and basin represented individual crustal blocks that, after being raised, had sunk to different levels as the crust was stretched, causing the blocks to tip slightly, like bergs of ice floundering in the sea. Gilbert gave a name to this topographically distinct area that covers most of Nevada and western Utah—the one that is used today by geologists and geographers and becomes evident to anyone who has spent seemingly endless hours driving the long, straight, almost always desolate but always majestic highways of this region of the continent: the Basin and Range Province.

Gilbert's view was a totally new way of looking at the origin of some mountain ranges. And his creative mind did not stop there. Looking through a telescope at craters of the moon and wondering whether they were huge volcanoes or scars of meteor impacts—most scientists then favored a volcanic origin—Gilbert did his own experiments and threw balls of hard clay against slabs of wet clay, deciding, from a similarity in shape, that lunar craters had an impact origin. He also spent years studying the Great Salt Lake and concluded that what he was seeing was a small remnant of a previously vast body of water, which he named Lake Bonneville after an early explorer of the American West. From Gilbert's work, we now know that this prehistoric lake once covered a third of the state of Utah, was more than 1,000 feet deep, and had emptied northward into the drainage of the Columbia River, leaving a very salty shadow of its former self behind thousands of years later. It was during one of his summertime trips to uncover more evidence for this ancient lake that he was on the western edge of the Basin and Range Province and decided to visit the site of the 1872 earthquake in Owens Valley After only a week of fieldwork, he was ready to propose a new theory of earthquakes.

Gilbert immediately realized that the key to understanding exactly what caused earthquakes was the long scarp that had formed during the earthquake. Unlike cracks and fissures that usually follow topographic contours, the scarp cut across streambeds and through hills, indicating that something deep in the Earth had controlled where it had formed. More important, as Gilbert studied the face of the scarp he noticed he could match features on one side of the scarp with features on the other side if the ground surface had slid as much as 15 feet *horizontally*!

Think of it this way: Take a newspaper column and rip it lengthwise along a straight line. Now shift the two parts and place them against each other. You can tell the amount of the shift by matching original lines of type. In the same way, by matching arrangements of rocks and other subtle features—known today in geology as finding the "piercing points"—Gilbert could tell that the ground surface had shifted horizontally along the scarp. And that was the great realization: Ground shaking and the pull of gravity would have caused a downward settling, like shaking a bag of rocks, and so motion along the scarp, if it was a secondary feature, should have been vertical. But the markings on the scarp ran horizontal—and by as much as 15 feet. From that, Gilbert concluded that the formation of the scarp was not an *effect* of the 1872 earthquake, but the *cause*. The Earth's surface wasn't moving up and down but side to side. But why?

Gilbert hypothesized the following: Imagine, he said, that you are in a railway car and the brake is set. Then if the car is being pushed or pulled, at first the car remains stationary, held in place by friction of the iron wheels against the rails. But eventually the pushing or the pulling becomes too great and the wheels slide a short distance along the rails, causing the entire railway car to shake momentarily.

According to Gilbert, a similar thing happens inside the Earth. In that case, some force, some impulse—which would not be known for another 80 years—causes strain to build up within the Earth's crust. That strain is released—like between the wheels and the rails—when crustal blocks slide against each other. And because there is friction between the blocks, seismic waves are produced.

Today this all seems so obvious, but in 1883, when Gilbert did his work in Owens Valley, it was revolutionary. And, as often happens with a new idea, it was ignored—in this case for more than 20 years.

Not until 1906, when an earthquake devastated San Francisco and the surrounding area, did geologists readily accept Gilbert's idea that earthquakes were caused by the sliding of crustal blocks. That earthquake, as will be shown, gave undeniable proof that the sliding of crustal blocks had caused the shaking. Barely a decade before then, someone had discovered the San Andreas Fault.

CHAPTER 2

NO OCCASION FOR ALARM

"You, you I want," said the earthquake.
"Not yet," said San Francisco.
—*Sacramento Bee,* March 1, 1904

For centuries, the men of Anstruther, Scotland, lived most of their lives out at sea. That would have been the fate of Andrew Lawson, born in Anstruther in 1861, but his father suffered a shipwreck—an ordeal that lasted two weeks and that left his hearing impaired and his heart weakened—and Andrew's mother took charge of the young family. She moved them to Hamilton, Ontario, for no other reason, she would say, than to prevent her son from going to sea.

Hamilton has a peculiar feature running through it—a giant step a few hundred feet high that divides the city in half. If one follows the step, one finds that it begins in northwest New York, passes through Ontario, then swings through Michigan and Wisconsin. Known as the Niagara Escarpment, it forms the cliff over which water cascades at Niagara Falls. Geologic work has revealed that it is an erosional feature that follows the former shoreline of an ancient tropical sea. For this story, the importance of the escarpment is what is found in the vertical wall—an abundance of fossils from the Ordovician and Silurian periods—that is, from about 500 to 400 million years ago.

These are especially well exposed in Hamilton, where one can find fossils of sponges, brachiopods, and crinoids, the last a marine animal that had a stalk and a feathery top that made it look like a miniature palm tree. Such natural oddities that are easy to collect attracted the interest of local children, and young Lawson was no exception, though in his case the fascination sent him on a lifelong journey.

In 1883 he graduated with honors in natural science from the University of Toronto. That led immediately to a job with the Geological Survey of Canada, which assigned him to lead a field party on a survey of the Lake of the Woods region north of Lake Superior. The purpose of the survey was to decide where mineral ores might be found. The most knowledgeable geologists and the most experienced miners had already determined that such economically important deposits must exist—it was just a matter of finding them. But, true to his nature, Lawson came to challenge their most basic assessment.

After three summers of work, following canoe routes through Lake of the Woods and studying every rocky outcrop he could find, Lawson concluded that the vast underlying mass of rock had been misidentified. And the mistake had been made by none other than Sir William Edmond Logan, the first director of the Geological Survey of Canada and the person for whom Mount Logan, the highest peak in Canada, is named.

Logan had proclaimed the rock to be old sediments that had been altered by heat and pressure. But Lawson saw it as granite—admittedly a very old granite, one that had been deeply weathered because of its great age, but granite all the same.

Lawson wrote his report challenging conventional opinion and his superiors blue-lined it, telling him to change major parts. Instead, he took the report to the queen's printers and told them to publish the report as originally written. Though decades of subsequent work would show that Lawson was correct, this was not the way to advance one's career. In 1890, Lawson resigned.

He found work as a consulting geologist in British Columbia, hired to study the coalfields at Nanaimo next to Queen Charlotte Sound. That provided him ample opportunities to explore the nearby valleys and inlets by train and ferry. After one of these sojourns, when he returned to Vancouver, where he lived, a letter was waiting for him. It was from California and it offered him a professorship at the University of California, which then consisted of one campus located in Berkeley.

Joseph LeConte, head of geology at the University of California and soon to be a co-founder of the Sierra Club with John Muir, never adequately explained why he hired Lawson, except to say that Lawson would teach students the science of geology while he would instruct them on the philosophical implications of the science.

The two men had met a few years earlier at a science conference in Canada, so LeConte was probably aware of Lawson's difficulties with his superiors. From a hint given in his later writing, it is also known that Lawson was hoping to find work and move to a warmer climate.

California could provide that—and much more. Essentially, nothing had been done to advance knowledge of the geology of the state since Whitney and his few assistants had left and the California Geological Society had been abandoned in 1872. Almost anything Lawson chose to do would be original. And because so much of the land was well exposed, he would have the further benefit of being able to follow geologic units over long distances, something that was impossible in the heavily forested Lake of the Woods or in British Columbia.

Lawson arrived at the university in October 1891. By the next summer, he was ready to engage in his first program of geologic research in California. He chose the Coast Ranges south of San Francisco because they were close and because, in a small corner, they contained a geologic deposit of considerable economic importance. A quicksilver mine near the southern end of San Francisco Bay that had been in operation since the beginning of the Gold Rush was the only known source of cinnabar—a mercury sulfide—in the state. Mercury was essential to extract gold from ore, so it was obvious that a component so crucial to California's gold-mining industry should be investigated.

Lawson enlisted the aid of a promising student, Charles Palache, to work as his field assistant. The selection was a good one: Palache would go on to be a distinguished geology professor at Harvard, and while at Harvard would assemble one of the most impressive mineral collections in the world. But for now he was a student, which meant Lawson sent him ahead to make preparations for a summer of geologic fieldwork.

Palache procured a horse, a wagon, and supplies in San Francisco and drove the wagon to Colma, at the end of a train line that ran south of the city. On July 7, 1892, Lawson arrived by train. The next day, a Friday, the two men began what would be a momentous two weeks that ended not only in the discovery of the San Andreas Fault but, using a keen sense of geologic cunning, also led Lawson to realize that the fault had even slid recently and therefore must still be active.

The first indication that there might be something of particular geologic interest in this region of California was the existence of a remarkably straight and deeply trenched valley that does not run parallel to the mountain ridges of the Coast Ranges but rather cuts across them. This feature is the San Andreas Valley, named by a Spanish expedition that camped within it on November 30, 1774, the feast day of the apostle known to English speakers as Saint Andrew. In the 1860s, the city of San Francisco took advantage of the large size and proximity of this extraordinary valley to pay the Spring Water Valley Company to hire more than 300 Chinese laborers to construct a dam to impound water within the valley. A long pipe was laid, and through it reservoir water was sent to San Francisco to be used by the city's businesses and private homes and by members of the municipal fire department to fight any conflagrations.

The San Andreas Valley is readily seen by anyone who flies into or out of San Francisco International Airport. The airport is located on what was once a marshy plain at the edge of San Francisco Bay. To the north is an isolated block that forms San Bruno Mountain. Colma is immediately west of the block. Three miles south of Colma is where the San Andreas Valley begins.

The water reservoir lies along the axis of the valley. If one projects the axis northward, it intersects the Pacific coast at Mussel Rock. It was here that Lawson and Palache found the next piece of evidence that drew their attention to the importance of the valley.

For the next few days, they did reconnaissance work along a wagon road that ran from Colma to the coast. North of Mussel Rock they found a thick section of loose marine sediments; from Mussel Rock south was a system of hard rocks—greenstones and granites. Such a juxtaposition of different rock types always piques a geologist's interest and cries out for explanation. After just a single day, Lawson and Palache could provide one.

On the morning of Wednesday, July 13, Lawson and Palache took a stroll, one that can be easily replicated today, along the long stretch of sandy beach north of Mussel Rock, outlined on the landward side by a high sea cliff. I suggest making the descent to the beach at Thornton Beach State Park, a popular site for paragliders, in Daly City. From there, one heads south.

The layered marine sediments exposed in the sea cliff are rocks of the Merced Formation—named by Lawson—comprised, as he wrote in his field notebook, of a "fine section of fossiliferous sandy clays, soft sandstones, and shell beds cemented hard." What is important here is how much these sedimentary beds, originally laid down as horizontal layers, have been tilted and now deviate from a horizontal line.

For the first few miles south of Thornton Beach, the beds of the Merced Formation are nearly horizontal, but that changes as one nears Mussel Rock.

As Lawson recorded in his notebook, within a quarter mile of Mussel Rock (the haystack-shaped greenstone sentinel that sits just off the coast in the zone of breakers), the beds of the Merced Formation are tilted at various angles, some as steep as 65 degrees. Soon after, he offered an explanation, declaring that "a great fault" existed between the soft rocks of the Merced Formation and the hard greenstones and granites south of Mussel Rock, and it was movement along this fault that had caused the normally horizontal beds of the Merced Formation to become jumbled. Furthermore, it was no coincidence that Mussel Rock lay along a projection of the axis of the San Andreas Valley; in fact, the fault probably extended along the length of the valley, a distance of 40 miles. And it was this association that led Lawson to give it a name, the San Andreas Fault, the name first used in a government report prepared by Lawson and published in 1895 that described the geology of the San Francisco peninsula. But it was long before this that Palache made a key observation that quickly led Lawson to conclude that the San Andreas Fault had recently moved.

The day after they strolled on the beach, Lawson left Palache alone for a few days while he returned to San Francisco and Berkeley to attend university meetings and to conduct university business. But he gave his student instructions: Lawson wanted Palache to search for outcrops of rocks that might exist on the east side of the newly discovered fault.

Palache did as he was told. On July 20, he began at the edge of the sea cliff in what is today the Westlake District of Daly City. There the deeply soil-covered surface was planted in vast fields of cabbage cultivated right up to the edge of the cliff. Palache rode his horse through the fields, then through pastures where cows grazed. He was unable to find any outcrops, but he did find something curious.

Little ponds were sometimes midway up a slope or atop ridges, in places that seemed unlikely for ponds to exist. They were long and narrow and had neither outlets nor streams feeding them. Most curious of all, they were strung along a long line.

Palache was puzzled by them but saw no special significance in their form and location. He of course mentioned them to Lawson the next day and was surprised when the professor became excited and insisted that he be shown the ponds at once.

None of the ponds exist today, though by using old maps a few of the depressions where the ponds once lay can be located. Some are now small community parks, such as Imperial Park in Pacifica and Callan Park in South San Francisco. One of the largest ponds has been filled and a softball field built over it at what is today an athletic complex known as Fairmont Park, also in Pacifica. One wonders, watching players hit balls and race around the bases, whether any of them knows what lurks beneath the ground.

Lawson knew. And he realized this not because of what he saw but because of what he did *not* see.

Geology is about sleuthing. And to be a good field geologist is to be an exceptional detective. Here one is reminded of the famous Sherlock Holmes mystery "The Adventure of Silver Blaze" in which the crucial clue is a watchdog that did not bark—a story, by interesting coincidence, first published in December 1892, a few months after Lawson discovered and first studied the San Andreas Fault. And what Lawson did not see along the line of small ponds in 1892 was the geologic equivalent of not hearing a dog bark.

The land west and south of San Francisco was still mostly vacant in the 1890s, much of it covered by sand dunes that shifted during strong ocean breezes. As the sand shifted, some of it would get caught in and fill depressions. The fact that several small ponds existed in a small area near the coast west of Colma was evidence that the depressions that held those ponds must have formed recently.* Thus,

* Such local depressions, which are now known as *sag ponds,* are commonly recognized along active faults and indicate a place where there is an offset along a fault strand where earthquake movement is stretching the land, thus causing the ground to drop and form depressions. Conversely, in some places earthquake movement can produce a compression, in which case the land is pushed up and forms what are termed *pressure ridges.*

because these particular ponds lined up along an extension of the San Andreas Valley where Lawson just a week earlier had determined a fault existed, from the alignment of ponds and, hence, from the alignment of depressions, he concluded "movement along this fault zone is still in progress."

And, indeed, it is.

But Lawson left out an element that we would consider essential today: Does the segment of the San Andreas Fault that he discovered in 1892 represent a seismic risk?

The question probably did not even occur to him. The idea that the slippage of geologic faults was the cause of earthquakes had been proposed only eight years earlier—by Gilbert, after his study of the 1872 Owens Valley earthquake—and it was an idea that was not yet widely accepted. How Lawson felt about the matter goes unrecorded, because few geologists—Gilbert being an exception—considered it relevant to their science.

In fact, the scientific study of earthquakes had begun only a few decades earlier, and that study was not undertaken by someone even remotely connected to geology. Instead, initially it was pursued by a man who had a more practical concern—the construction of buildings.

Robert Mallet, an Irish engineer, became interested in earthquakes because of a diagram he saw in *Principles of Geology*, first published in 1830 and written by Charles Lyell—the same man who influenced Whitney—that showed how the passage of waves of a large earthquake in Italy in 1783 had caused heavy stones stacked to form a tall obelisk to rotate. Lyell gave this as evidence that, in addition to the back-and-forth and up-and-down motions of an earthquake, there was also a rotational, or "vorticose" one. Mallet, an accomplished designer and builder of major structures—his constructions include the terminal station of the Dublin and Drogheda

Railway, a large polygonal shed with a giant hydraulic turntable at the center; a series of swivel bridges across the River Shannon; the Westminster Bridge in London; the Fastnet Rock lighthouse that stands on the southernmost point of Ireland and thus was the last part of Ireland seen by emigrants sailing to the United States—was well practiced in the art of moving and positioning heavy objects, so he knew that a stone could be rotated simply by applying an appropriately directed force.

A rotation came about if the geometric center of the base of a heavy stone did not lie directly beneath the center of gravity of the stone. From the diagram shown by Lyell, Mallet saw that was the case for the stones that formed the obelisk.

That simple realization led him on a lifelong quest to understand earthquakes based on his practical knowledge of mechanics. In 1857, with letters of support from Charles Lyell and from Charles Darwin, he received funds from the Royal Society of London to go to Italy and study the devastating effects of a recent earthquake.

At first, he admitted, he found himself "in the midst of utter confusion." But with the practiced eye of an engineer, he soon saw a pattern to the destruction. There was a central area, measuring a few miles across, where almost every building had collapsed and where there had been the greatest loss of life. Beyond this was a much larger area where many people had been injured and where many, though not all, buildings had collapsed. He identified a third area, larger still, where a few buildings had fallen and where there had been almost no loss of life. And there was a fourth area, extending for hundreds of miles beyond the central area, where almost everyone had felt the earthquake but there was little to no visible damage.

What he produced was the first isoseismal map of an earthquake, something that is now routine and used by engineers to gauge the level of destruction and by scientists to estimate the intensity of an earthquake.

Mallet also had a considerable knowledge of explosives. His expertise was so well regarded that in 1835 when the water well used by Guinness and Company in Dublin to produce its ale gave out, Mallet was hired to restore it. He did so by boring a three-inch-diameter hole deep into the solid rock at the base of the well, then packing the hole with gunpowder and exploding it. The shock shattered enough rock that the well has never been dry since.

Mallet now applied his knowledge of explosives to determine the speed of seismic waves. He began by burying 25 pounds of gunpowder at a sandy beach in Dublin County. Then, from half a mile away, he watched. When the gunpowder exploded, he turned his attention to a saucer filled with mercury, recording how long it took for the first ripples to appear. From this beginning, his experiments grew to 12,000 pounds of gunpowder exploded at a rock quarry in Wales. Eventually he concluded—and it was the first time anyone had done so—that waves produced by earthquakes travel at speeds of more than a mile a second through the Earth, a phenomenal rate when one considers that the fastest mode of transportation of the era was by train, which then traveled at speeds of no more than a few miles per hour.

In 1852 he felt his first earthquake, which he remembered as a "heavy thump" that woke him from sleep and which he did not realize was an earthquake—initially, he thought it was a burglar breaking into his house—until he talked to neighbors the next morning. Probably of no connection to this singular event, the same year, Mallet began his greatest undertaking—to compile a catalogue of all major earthquakes that had occurred in history.

His effort required several years, and he published his catalogue in several installments. In the fourth installment, published in 1858, he coined the word "seismology" to describe this and his other work on earthquakes.

Mallet's catalogue in its completed form contained 6,831 events, giving the date, the location, a brief description, and the source. The first entry in the catalogue is the earthquake noted by Moses when he received the Law on Mount Sinai and, according to Exodus 19:18, ". . . the whole mount quaked greatly." The first nonbiblical earthquake comes from a source quoted by Aristotle, which Mallet lists as occurring in 1450 B.C. when in central Italy "a city was swallowed up, and a lake produced in its place." The earliest event listed that occurred outside Europe was in China in 175 B.C.

Notwithstanding the questionable reliability of some of the early entries, Mallet's catalogue—and its many descendants—has proven to be a valuable resource for scholars working in other fields. For example, there has been a long-standing question as to when Shakespeare wrote one of his most famous plays, *Romeo and Juliet*. The answer, it seems, can be found in Mallet's earthquake catalogue.

In Act I, Scene 3, the Nurse makes the statement "Tis since the earthquake now eleven years." Most scholars took this to be a reference to an earthquake felt throughout England, especially at London, Dover, and the whole of Kent, late on the afternoon of April 6, 1580. Shakespeare undoubtedly felt it. And so, it was reasoned, this was the earthquake that prompted the comment by the Nurse. If so, then *Romeo and Juliet* must have been written in 1591, but those who are expert on the bard's changing style say this is at least a few years too early.

Mallet's catalogue suggests another possibility. An earthquake on the afternoon of March 1, 1584, was felt strongly throughout Switzerland, eastern France, and northern Italy, the last including Verona, where the play is staged—but not in England. William Covell, an English clergyman and critic who often praised Shakespeare's works, was in northern Italy at the time of the earthquake, describing the event as "the terrible Earthquake" that "moved the lesser plants," and it might have been he who described the event to Shakespeare, who then incorporated it into the play he was writing at the time. In this case, *Romeo and Juliet* must have been written in 1595, a year that does conform to the bard's style of the time and is two years before the first publication of the play, which was in 1597.

Mallet did not compile the first earthquake catalogue—that distinction belongs to Florentine humanist Giannozzo Manetti who, in 1457, published a catalogue of 70 earthquakes in Italy that occurred from antiquity to the mid-15th century—but at the time, Mallet's work was by far the most extensive. And then he did something that no one had yet done—he used it to make a map of worldwide seismicity.

Instead of plotting the location of individual earthquakes, which is the standard today, he applied three different shades of color to the map to distinguish the intensity of activity. The lightest shade showed where only minor shocks occurred that did not produce serious commotion or destruction; the darkest shade indicated where earthquakes were numerous and violent. Looking at the general pattern, Mallet got it exactly right.

Mallet identified two great belts of seismicity. One, in his words, followed "a vast loop or band round the Pacific," while the other was a broad zone that ran from the East Indies across southern Asia through the Middle East and the Mediterranean. The "loop" is today's familiar Pacific Ring of Fire, where not only earthquakes but volcanic eruptions are common. The broad zone is where mountain masses are forming—the Himalayas, the Zagros of Iran, the Taurus of Turkey, the Apennines of Italy, and the Atlas of northern Morocco. Today we know that 90% of all the seismic energy released inside the Earth occurs along these two belts. It is the details that one quibbles with Mallet.

Mallet's map equates the level of seismic activity in California and in New England, identifying both as "minor." It is an idea that did not last long. In the 1860s, the decade after Mallet published his map, two damaging earthquakes struck near San Francisco, and California was on its way to becoming known as earthquake country. But if one consults his catalogue—and the catalogues published later—one finds that the most intense seismic activity to strike the United States in the 19th century did not occur in California—or along either coast—but at midcontinent, close to where the Missouri, the Mississippi, the Ohio, and the Tennessee Rivers come together.

According to Mallet's catalogue, on December 16, 1811, "the disturbance of this region now commenced." The initial shock, which occurred early in the morning, was felt over most of the eastern United States. In Washington, D.C., Dolley Madison, the president's wife, felt the shaking and wrote of it to her sister. In Pensacola, Florida, people were awakened when they heard their houses crack, then watched as doors and window shutters moved. In Springfield, Illinois, the shaking was strong enough to stop clocks. In Nashville, some chimneys were thrown down. And in the frontier town of New Madrid in southeast Missouri, trees shook so violently that branches snapped and fell to the ground.

The activity persisted, the people of New Madrid feeling individual shocks daily, sometimes hourly, for months. The next violent shock came on January 6, 1812, when, according to Mallet, "the town of New Madrid was greatly injured." This earthquake too was felt across the country.

John James Audubon was riding a horse through west Kentucky when the earthquake hit. He heard a distant rumbling and thought it was a tornado. He tried to spur his horse, but the animal refused to move. He dismounted just at the time when all the shrubs and trees began to move and the ground rose and fell "like the ruffled waters of a lake." He knew it was an earthquake from descriptions he had read but, as he would reflect, "What is a description compared with reality?" The shaking lasted about a minute. Then, everything calm, the horse responded again to his commands and he continued on his travels.

The most severe event of the series, which quickly became known as the "hard shock" by the people of New Madrid, came early on the morning of February 7. A boatman named Matthias Speed had tied his boat to a willow bar on the Mississippi River near New Madrid the previous night. At about three o'clock, he was awakened by the violent agitation of the boat. To his shock, he came to realize that not only was the water rising, but the Mississippi River was flowing the wrong way!

It took an entire day for the river to resume its normal course and for the water to drop to the level of the previous day. By then—and Matthias Speed was one of the few who witnessed its formation—a new lake, 15 miles long and 5 miles wide, known today as Reelfoot Lake, had formed on the east side of the Mississippi River at the base of the Chickasaw Bluffs in western Tennessee. The stumps and roots of trees killed by the sudden drowning can still be found projecting above the lake surface in shallower areas.

Such stories—and they were many because, as Mallet knew, earthquakes are common in the United States—showed that the people of the United States had a knowledge of such events. But they did not consider *their* earthquakes to be destructive, unlike the ones that happened in places like Lisbon in 1755 or seemed to strike Japan repeatedly. That is, not until 1886, when the first major seismic disaster hit the country. And it did not occur in California, but in a place that few people today associate with earthquakes—South Carolina.

At ten minutes to ten o'clock on the Sunday night of August 31, 1886, a gentle vibration was felt by many of those who were still awake in Charleston, South Carolina. A young boy who 20 years later would find himself in San Francisco to experience that city's great earthquake, recalled that this initial vibration reminded him "of when a cat trots across the floor." But the slightness of the disturbance was short-lived. Within seconds, the house he was inside was shaken by a sharp jolt. He watched out a window as people, frantic with terror, rushed into the streets, some of them crushed and later dying after chimneys and brick walls fell on them. Then, as the boy, now a man, recalled years later, there was "a final mighty wrench" and the shaking was over almost as suddenly as it had begun.

The city was left with many of its wood-framed houses with a permanent tilt, a feature that became known as the "Charleston lean." Some of these houses can still be seen today and can be found in the historic district of Charleston. Inside, new floors have been constructed over the old inclined ones.

As expected, the city suffered a setback. Not a single building in Charleston escaped unscathed—and most were seriously damaged. The official number of fatalities was initially reported as 60, though later estimates put it at twice that number.

It was the most destructive earthquake yet recorded in the eastern United States. More than half the population of the country felt the shaking, with reports coming from as far away as Davenport, Iowa; Green Bay, Wisconsin; and Burlington on Lake Champlain in Vermont.

The distant shaking made it possible to make an important scientific measurement—the first of its kind for an earthquake. In 1883, just three years before the quake, to satisfy a need for train conductors to have synchronized watches so that they could avoid the collision of train engines—a deadly collision had already happened in New England—standardized time zones were established. And so, at the time of the Charleston earthquake, many households and businesses across the country and many men had their clocks and pocket watches set precisely. That meant that when the seismic wave spread across the country, stopping pendulum clocks and causing men to look at their watches, dozens of good time records were available. From this, it was an easy matter to compute the average speed the earthquake shock waves traveled, which was determined to be *three and one-fourth miles per second,* confirming Mallet's early estimate.

Neither the events at New Madrid nor the Charleston disaster led to any immediate interest in earthquakes, probably because neither area had persistent seismic activity. That was not true of California, where, since the mid-1860s, it was recognized that such activity was frequent—certainly higher than the "minor" designation given by Mallet when he published his map of global seismicity in 1862. If one delved into the early history of the region, one would find that the first person who traveled in what is today the state of California and wrote about it—that person felt not one, but *several* earthquakes.

In the summer of 1769, perceiving a threat from the Russians to the north and the British to the east, the government in Spain sent out an expedition from Mexico to secure the land along the Pacific coast as far north as Monterey Bay. (San Francisco Bay was still not yet known to anyone except local Native Americans.) Accompanying the expedition was the Franciscan Fray Juan Crespí.

On July 28, two weeks into the journey, the expedition was camped along the Santa Ana River near present-day Anaheim when, according to the journal kept by Fray Crespí, a sharp earthquake was felt that "lasted about half as long as an Ave Maria," that is, about six seconds. Three more earthquakes were felt later the same day.

The expedition continued north, Fray Crespí recording at least one earthquake each day. On one day, he recorded twelve events. The shaking of some of them, so he wrote, was violent.

By August 3, the expedition was in the San Fernando Valley. And no more earthquakes are recorded in the Franciscan's journal.

What, then, might have been the reason for the flurry of earthquakes? The best guess is that this first-ever expedition to California happened to be traveling through an area where a moderate earthquake had struck recently, probably within a few days or weeks. So the earthquakes they experienced were probably aftershocks. A comparison with more recent events suggests the supposed moderate earthquake that preceded the flurry was an event similar in size and in location to the 1933 Long Beach and the 1987 Whittier Narrows earthquakes. The fact that Fray Crespí was in the San Fernando Valley and that he recorded no earthquakes after August 3 suggests he was too far north to feel additional events.

The first attempt to maintain a list of California earthquakes was made in 1856 by John Trask, one of the founders of the California Academy of Sciences, who had come west with John Woodhouse Audubon, son of the famous ornithologist (who had felt the shaking of at least one of the New Madrid earthquakes), via an overland trail through Texas and northern Mexico, arriving in San Diego on November 4, 1849. The two men went immediately to the gold diggings in the Sierra Nevada but, finding the work of mining exceptionally hard, their fellow miners violent and unsavory, and accumulating no immediate wealth, Audubon returned to the east and Trask settled in San Francisco, where he resumed his previous profession as a physician. From that, he met other learned men who had also come to California to search for gold, then in desperation had restarted their professional lives. This coterie of educated men, including Trask, began the academy in 1853. Trask took the title of "curator of geology." His job was to collect fossils and uncover facts about California's earthquakes.

He compiled information about earthquakes from newspaper and telegraphic reports and by interviewing other residents of the state. His first catalogue of California earthquakes—Trask would publish annual addenda—included 59 events that had occurred since 1850. The first event listed was on March 12 of that year when "a light shock was felt in San Jose." On May 15, 1851, there were "three severe shocks in San Francisco," the shaking strong enough to break some windows and throw merchandise from shelves to the floor of a store on California Street.

A reading of his catalogue shows that there was an earthquake felt somewhere in California almost every month. Furthermore, though more earthquakes were recorded in the northern half of the state, where most of the state's population resided, it was in the southland that earthquakes seemed to be more severe. On November 26, 1852, an "earthquake was felt over the entire country east and south of San Luis Obispo, to San Diego and the Colorado river." The idea that earthquakes were more severe in the southern half of California was reinforced when, on the morning of January 9, 1857, shaking was felt across almost the entire state. The tremor was centered in the mountains north of Los Angeles, where one, possibly two, people died.

A significant amount of shaking of that earthquake was also felt in San Francisco, where, at a quarter past eight in the morning, three printers for the *Daily Evening Bulletin* were at work on the third story of a building on Merchant Street. As one of the printers would recount, the floor they were standing on started to tremble and move. He grabbed his coat and prepared to run. Another printer was "in such great tribulation" that he froze when he could not find his hat. The third, "the fat man," resigned himself that there was no chance for escape. He held on to a printing case, perfectly resigned to his expected fate. But the shaking did stop after a few seconds, and the three men resumed work, noticing that the pendulum clock in their office had stopped.

This story of the three printers illustrates that as far as most people in San Francisco were concerned, earthquakes near their city were a novelty. There was no record of major damage—nor a fatality—in the short history since the beginning of the Gold Rush in 1850. And Trask's catalogue confirmed it. But that impression changed in the mid-1860s when two damaging earthquakes struck.

One of the city's newest residents, Samuel Clemens, who had recently begun to use the pen name "Mark Twain," was walking down Third Street when the commotion started. It was the early afternoon of the Sabbath and the only two objects in motion anywhere in sight were a man in a buggy behind him and a streetcar making its way slowly up a cross street.

Mr. Clemens heard a commotion and, thinking it was a fight and wishing to see it, broke into a run and turned a corner, around a framed house, just in time to hear a great rattle and feel a sudden jar. Then there came a terrific shock.

"The ground seemed to roll under me in waves," he wrote of the new experience, "interrupted by a violent joggling up and down, and there was a heavy grinding noise as of brick houses rubbing together."

He reeled about on the pavement, trying to keep his footing. He pushed up against the framed house, hurting his elbow. And then the shaking intensified.

He watched as the entire front of a tall four-story brick building sprang outward like a door and fell across the street, raising a great volume of dust. Then the buggy came by and the driver was thrown out and the buggy ended up distributed as small fragments for hundreds of yards along the street. The horses pulling the streetcar had stopped and were rearing. The passengers were pouring out of both ends, one man crashing through a glass window to escape.

And then, from every door to every house as far as the eye could see, people streamed into the street. Never had he seen a solemn solitude turn into teeming life in so fast a fashion.

It was October 8, 1865. From the many eyewitness accounts of shaking and destruction, the event originated north of Santa Cruz, either on or near the San Andreas Fault.

Another destructive earthquake came soon after, again occurring in October, which led to speculation that earthquakes near San Francisco were influenced by the seasons or by a particular pattern of weather. As will be seen, both ideas were soon discounted.

The earthquake of October 21, 1868, originated on the east side of San Francisco, and it left a ground rupture running for 20 miles southeast of San Leandro, along which the surface was shifted as much as three feet. Thirty people were killed during the shaking, five of them in San Francisco. All deaths were caused by the collapse of buildings.

This should have been a wake-up call to Californians that powerful earthquakes could strike their state and do considerable damage—but the call was not heeded. In fact, the possible threat of future earthquakes was ignored.

Members of the California Academy of Sciences did prepare a report about the 1868 earthquake and, though there is some confusion on this point, there may have been an attempt to distribute it, but the report was suppressed. Exactly by whom and whether it was an organized effort is unknown. But no trace of the report has ever been found.

And yet California continued to be regarded as earthquake country, and those who lived in the state and traveled outside often touted themselves as experts. A case in point was California senator William Gwin, who happened to be in New York City when that city was struck by what is still the largest earthquake in the city's history.

It was midafternoon on Sunday, August 10, 1884. The effects were felt from Richmond, Virginia, to Portland, Maine. An indication of the severity of the shaking came from Hoboken, New Jersey, where a 300-pound man, asleep on his bed, was immediately thrown to the floor. Two police officers standing at midspan on the recently completed Brooklyn Bridge reported that the great stone towers at either end of the bridge oscillated visibly several times, the bridge responding as if it had suddenly been struck by a hurricane. And in Brooklyn, John O'Mara of the fire department, who was on duty at the top of City Hall—his job was to spot fires—said that when the building started to sway, he was preparing himself from a quick ride down to the streets below.

Immediately after this was over, a reporter for the *New York Times* spotted Senator Gwin standing in front of a hotel. The reporter asked Gwin, "What on earth was that?"

The senator responded with no great enthusiasm. "That was an earthquake. I was raised on 'em."

"Did you consider this a severe shock?" queried the reporter.

"Yes, it was quite severe, but in California, I've been through all the earthquakes since 1849. The severest one was in 1868. Since then we have experienced repeated shocks, but all have been light."

"How long do you think today's shock lasted?"

"Not over five seconds," the senator replied. "To persons not used to earthquakes, it probably seemed much longer."

Then just before he left, the senator from California turned to the reporter and said, "Oh, yes, I've been raised on earthquakes."

Three types of people live in California: those who have never experienced an earthquake but hope to feel one; those who have felt a moderate amount of shaking and were intrigued by the ordeal; and those who have felt strong shaking. Ask anyone from the last group what it was like and that person will begin by staring at you in silence, then in deliberate tones take several minutes to recount exactly what they saw, felt, heard, and thought during what was probably only ten or twelve seconds of unbelievable confusion and mayhem.

For the first 12 years that Andrew Lawson lived in California, he was a member of the first group—his frequent travels taking him often outside the state. The first opportunity to join the second group came in 1898 when an earthquake originated just north of Berkeley. Two railroad bridges were damaged and several buildings in Vallejo, just north of Berkeley, collapsed. Lawson, however, was on a yearlong sabbatical in Europe when that earthquake happened. It was March 30, 1898, and the shaking overturned oil lamps, starting a half-dozen fires in Vallejo. It took a day to bring the fires under control, a harbinger of what was to happen in San Francisco in eight years.

Lawson was in Berkeley in 1904 when dozens of small earthquakes rocked the area for more than a month. The series began with a mild earthquake on November 27, then four quick ones, in less than ten minutes, on December 1. Those four were enough to cause chandeliers to swing and a few brick chimneys to collapse in Berkeley, Oakland, and San Francisco, the three largest cities in the state.

The main shock occurred at midmonth and it prompted a newspaper reporter to seek out and question Lawson about this peculiar sequence of events.

"History and records show that earthquakes in this locality have never been of a very violent nature," the professor said in what can only be imagined was an authoritative tone, "and so far as I can judge from the nature of the recent disturbances and from accounts of past occurrences, there is no occasion for alarm at the present time."

Two weeks after he had made that pronouncement, the last earthquake of the series happened, on New Year's Day, 1905. Newspapers wondered how many of the revelers even felt the event, though the shaking was enough to break some plate-glass windows and to dislodge one of the heavy iron pinnacles atop the City Hall on Larkin Street.

So once again seismic peace was the norm in San Francisco. But a certain irony creeps into the story here.

A year later, in January 1906, Lawson and former student Harold Fairbanks, the most recent graduate of the geology program at the university, took a trip to southern California to uncover information about the 1857 earthquake that had been felt across the entire state.

They started in the Cholame Valley of the Coast Ranges, 200 miles southeast of San Francisco. They talked to ranchers who had felt the earthquake. Some ranchers showed them the "earthquake crack" that had formed during the event.

Lawson left Fairbanks at midmonth and returned to Berkeley. Fairbanks continued south. With the help of local ranchers, he was able to follow the 1857 rupture through the San Emigdio Mountains north of Los Angeles and along the boundary defined by the Mojave Desert on the north and the San Gabriel Mountains on the south. As Fairbanks would soon write, the 1857 rupture had run along a "uniformly straight course" for more than 200 miles.

This realization—at this time—is remarkable because barely a month after he returned to Berkeley, a similar rupture, also uniformly straight in its course and also running for more than 200 miles, would form in northern California, abruptly changing the lives of Fairbanks and Lawson and the other half a million people who lived in and around San Francisco.

CHAPTER 3

A TUMULT OF MOTIONS AND NOISES

I stood in the doorway, as that was the safest place.

—Alice Eastwood, San Francisco, 1906

In 1906, near the end of a career that others would later characterize as "geologizing" the American West, Grove Karl Gilbert wrote: "It is the natural and legitimate ambition of a properly constituted geologist to see a glacier, witness an eruption, and feel an earthquake."

The glacier, so he said, came easy because it "is always ready." So too the eruption, which "has a course to run." But the earthquake, "unheralded and brief," so Gilbert wrote, might elude a geologist for an entire career.

His encounters with glaciers and eruptions had in fact come easy. In 1899, as a member of a scientific expedition to Alaska, he photographed and mapped more than 40 glaciers, at one point spending three days hiking across the giant rivers of ice that descended from mountains and into the upper end of Glacier Bay. On the same trip to Alaska, he sailed in a small ship within sight of Pavlov volcano—Gilbert was pleased to report that the volcano was erupting from its summit—then spent an afternoon climbing over the steaming crags of Bogoslof Island, which had recently risen from the sea. But the shaking of an earthquake still eluded him.

Gilbert left Alaska too soon to feel an earthquake that originated beneath Yakutat Bay and produced a 40-foot wave that swamped the shoreline. He had stood on the shoreline a month earlier and, when told of the shaking and of the wave, seemed genuinely disappointed to have missed both. Twenty-seven years earlier, he had been "tantalized," so he wrote, when on his first trip to the American West and after four months in California and Nevada, he had returned east just weeks before the 1872 earthquake in Owens Valley that was felt from Oregon to Mexico. But geologic upheaval finally found him. It was during the early morning hours of April 18, 1906, a Wednesday, and he was asleep in bed.

He had been living in Berkeley since the previous November, having come west at the request of President Theodore Roosevelt, who asked Gilbert to determine whether hydraulic mining should be resumed to extract gold from the hills of loose rock on the east side of the Sierra Nevada. In the end Gilbert would recommend against it, noting that hydraulic mining had a much wider impact than the silting of rivers. It had altered the flow of tidewater into and out of San Francisco Bay and had affected the shape of the tidal bar outside the Golden Gate, which meant it had a negative impact on shipping and commerce in the bay.

But this report was years in the future. For now, he was asleep in a private room on the second floor of the Faculty Club at the university—a place, he confided to a friend, "the typical Club-man would find dull," but which provided him with adequate comfort.

On the fateful morning, Gilbert was awakened by "a tumult of motions and noises." His immediate reaction was one of joyful anticipation because, as he would write, "It was with unalloyed pleasure that I became aware that a vigorous earthquake was in progress."

He soon realized that the noises were coming from the creaking of the building, made of a frame of heavy redwood, and from the rattling of furniture—both sounds diverting his attention from the specific sound associated with the earthquake itself. As he lay in bed the shaking intensified, and he noticed that although the motion came from many directions, the dominant one was a swaying in a north-south direction. Later, after the shaking had stopped, he watched the electric lamp suspended from the ceiling in his room continue to swing in a north-south direction and noted that water from a pitcher and liquid from a chamber pot had sloshed out to the south.

He rose immediately and inspected his surroundings. Except for the slight movement of furniture, nothing was displaced in his room. He searched other parts of the building and the outside of nearby buildings and saw no damage. Later, a thorough inspection of the university campus done by others would show that some books in the library had fallen off shelves, that a few chimneys had been twisted, and that some glassware in the chemistry lab had fallen and shattered on the floor. But there was no major damage evident in Berkeley.

After Gilbert left the Faculty Club, he walked the campus grounds. The sun had just risen and he would have noticed the promise of a warm spring day. At some point, perhaps late in the morning, he must have encountered Lawson, and the two men would have exchanged stories. Both had been awakened by the shaking. Lawson felt sure that there had been two distinct shocks separated by a brief lull. The second shock was the more violent of the two, and it had been strong enough to bring down a massive four-flue brick chimney atop his roof, the only damage to his house as far as he could determine. The debris, Lawson noted, fell to the southeast.

The two men probably sought out Armin Leuschner, a professor of astronomy who was in charge of the two seismographs at the university and also the two at Lick Observatory, 50 miles to the south. These instruments had been purchased to determine how earthquake shaking—even a slight amount of shaking—might affect the astronomical equipment at the university and at the observatory. So Leuschner was already knowledgeable about earthquakes—having experienced the 1898 shaking—and knew what one could do.

In his case, he had also been at home at the time of the severe shaking. And when it had started, he automatically did two things: He gathered his three children—ages one, four, and eight—in his arms and carried them outside, and as he did this, he also started counting to determine how long the shaking lasted.

Leuschner had even had the presence of mind—as Lawson had—to recognize that there had been two shocks and that the first one had lasted 40 seconds, followed by a comparative lull of ten seconds, then a second, stronger shaking that ended abruptly after 25 seconds.

As to the two seismographs at Berkeley, the arrival of the first earthquake wave had hit so hard that the recording drums and needles and driving clocks had been knocked into disarray and the records were useless. Leuschner spent much of the morning repairing and resetting them. For the next few days, the needles vibrated almost continuously, recording the occurrence of hundreds of aftershocks.

At this point, a question is raised: When did Gilbert or Lawson or Leuschner or any of the people they met that morning realize the severity of the earthquake and where it had originated? Lawson would remark that, from the lack of serious damage to buildings in Berkeley—he had gone off on foot and done a quick personal assessment of the city—initially he doubted whether it had been a significant earthquake. But there were two circumstantial reasons that were leading him to a different conclusion.

First, by midmorning, columns of smoke could be seen rising from the southern part of San Francisco across the bay. By mid-afternoon, these half-dozen or so columns had coalesced into a huge single billowing black cloud that towered over the city, the top of which was drifting eastward toward Berkeley. Detonations were also heard coming from San Francisco. But exactly what had happened was still a mystery.*

There were, of course, rumors coming from people crossing the ferries and landing in Oakland, but their stories did not have credibility and at this point had to be regarded as rumors. Obviously, the shaking had affected San Francisco and had done damage beyond. But how far?

And then there was the silence. All telephone lines and all but one telegraph wire to San Francisco had been cut off at the moment of the earthquake—and the single wire would be lost later that afternoon. Lawson's counterpart, geologist John Branner at Stanford University, located 30 miles south of San Francisco and where damage to buildings had been severe, would not be able to send out a message and describe his ordeal for two days. There was also no word coming from north of the city. In Santa Rosa and Petaluma, where some buildings were totally razed, a week would pass before reliable reports from creditable people were sent. The first indication that Lawson or Gilbert or Leuschner received that damage had been widespread and severe in places came to them later when a bicycle rider arrived from Lick Observatory.

Located atop Mount Hamilton east of San Jose, Lick Observatory was then the home of the world's second largest telescope and several smaller ones. After the earthquake hit, though there was only slight damage—none to the telescopes—the astronomers decided they should send word to the university in Berkeley that ran the observatory. They tried to use a telephone, but the connection was down. And so one of the night assistants—unfortunately,

* City officials had planned to fight any major fire in San Francisco with water piped from the reservoir in San Andreas Valley, but the earthquake had damaged the piping in several locations, namely where it ran across the San Andreas Fault, rendering the city without crucial water supply to fight the fires—or to supply drinking water after the disaster.

his name was never recorded—was dispatched by bicycle to place a call from San Jose.

It is 26 miles from Lick Observatory to San Jose. It is all downhill, but there are scores of tight turns. Upon reaching that community, the rider saw that all was confusion. Gas lines were broken and all but a few buildings had collapsed. There were no telephone lines or telegraph wires to the outside. So the rider got back on his bicycle and rode on.

It was another 50 miles to Berkeley. He passed through a string of small communities—Milpitas and Fremont, Hayward and San Leandro. In places, he had to dismount and walk his bicycle around gaping cracks in the road. If he had looked inside one of the many ruined houses, he would have seen that most of the plaster and almost all of the wooden laths and cornices had fallen and that loose articles had been violently thrown around. In Oakland, he would have noticed that almost all of the larger buildings, especially schools and churches, were totally wrecked. He reached Berkeley in the late afternoon and reported what he had seen.

Lawson responded by composing a note to the governor asking that the state of California formally establish a scientific commission to study what had happened. Leuschner remained with his seismographs, keeping them running, and recording whenever he felt the ground shake again. And Gilbert, if he had not yet done so, was now on his way to Oakland to try to find passage on a ferry to San Francisco.

It was mayhem at the pier. Refugees from San Francisco were already crowding the waterfront. Every place that could be used for shelter was filled. City officials had begun to lead processions of people inland, pressing into service all sorts of wheeled conveyances. Stray household pets, separated from their owners, seemed to be everywhere.

Gilbert tried to find passage to San Francisco but every captain refused, saying that soldiers on the other side would not let anyone disembark. So Gilbert left and returned to Berkeley.

He was back in Oakland the next morning. Across the bay, he could see that the entire waterfront of San Francisco was ablaze, with flames shooting up like discharges from a blast furnace. Gilbert could hear the distant rumble of dynamite, indicating that there was a desperate attempt to fight the fire. Reaching the Oakland pier, he saw that a sign had been erected: "Do not furnish passage to a single person to San Francisco until further notice. We cannot handle them."

And so, for a second time, he was unable to cross and returned to Berkeley.

A great earthquake had literally shaken him while he was in his bed, and now he was unable to study it. That frustration, certainly, played on his mind. But there was something else he wanted to see—a woman named Alice. She lived in San Francisco, where reports were now arriving of death and destruction and of people desperately fighting fires—and he must have wondered what had happened to her.

Born in Colorado, the first year she lived in California, 1893, Alice Eastwood made a solo ascent of Mount Shasta. She carried a backpack of her own design, which she had made by sewing two sheepskin rugs together. Inside was a bedroll, personal effects, and a small container of food—some sausages, canned sardines, a handful of figs, and a few slices of cheese.

She reached the summit at noon on the second day of her climb. Unfortunately, the view was marred by smoke from forest fires. She ate what remained of her food, not wanting to carry it down, and drank the last of her tea, which she had combined with a little whiskey. Then she began her descent.

Where possible, she slid down the mountainside. When she did so, she took out a burlap sack and cut two holes for her legs. She cinched the open end of the bag around her waist, tying it tightly with a cord. Then she jumped.

"I had a slight feeling of fear about the first effort," she would write, "but I had not more than commenced to slide when the fear vanished."

She laid on her back, bumping and gliding over slope after slope, her run finally ending thousands of feet down when she got into wet snow. Then she took to walking the remainder of the way, botanizing as she went.

Botany was the main passion in her life, followed by mountain climbing. So it was appropriate that she met Gilbert on an ascent of Mount Whitney in 1903, a climb sponsored and arranged by the Sierra Club, founded by John Muir and others in 1892 and of which Eastwood and Gilbert were early members. She was then curator of botany at the California Academy of Sciences and already held a large measure of fame for the many native plants of California she had collected and classified. Gilbert was traveling back and forth between the east and the west coasts and thus was able to see Alice only rarely.

He was a recent widower. She had had only one serious suitor in her life, a young man who had been too bashful to propose marriage. Gilbert and Eastwood tried to keep their romance private, but friends gossiped about them in letters and expected them to marry. In January 1906, for her 47th birthday, Eastwood received a card from Gilbert, who was then 62. Inside was a note. The note said that he realized she would always be devoted to her plants. It also said that he would "rather [she] be Alice Eastwood than queen or heiress."

On the morning of the earthquake, Eastwood was asleep in her third-floor loft apartment on Washington Street on the north side of Nob Hill. "The earthquake didn't frighten me," she later wrote, "as it was felt less where I live than in other parts of the city."*

---

* The severity of seismic shaking depends on how close someone is to the source of an earthquake and to the nature of the ground underneath. Eastwood was living on Nob Hill, which is solid rock. The people who lived south of Market Street felt stronger shaking because the ground beneath them was former marshland filled with loose debris.

At first she remained in bed, but as the shaking intensified, she rose and put on her bathrobe and stood in a doorway, as she had been instructed to do in case of earthquake by one of the other curators at the academy soon after she moved to California. At the end, a few dishes were broken, though only one prized one. She cleaned up the mess, ate breakfast, prepared her lunch, and started to walk to the academy, expecting this would be a normal day.

The academy building was on Market Street near the corner of Fourth Street, 12 blocks from her apartment. It was while she was walking down the steep south side of Nob Hill that she realized something serious had happened.

People were passing her walking up the hill carrying bundles wrapped in blankets. A few were dragging heavy trunks. She also noticed there was an unsettling quietness to the city. She did not hear the familiar clop of horses' hooves on cobblestones, and if anyone spoke it was always in a whisper.

The full scale of the calamity hit her when she reached Market Street. Electric and telephone wires were down. Not a pane of glass was left in any of the store windows. People were hurrying in every direction, some of them bleeding. And in the neighborhoods south of Market Street, which were then mostly wooden tenements, smoke clouds were rising, indicating that fires had already started.

Eastwood broke into a run. She reached the door of the academy building, but it was locked. There was an open doorway to the adjoining building where people were already taking things out. She entered there.

She climbed a few stories to where she knew there was a short bridge to the academy building, but the earthquake had collapsed the bridge so she went back to street level.

By now she could see a wall of flames a few blocks away. The academy director had arrived and unlocked the door and Eastwood raced inside.

The botanical collection was on the sixth floor. To reach it, one climbed a central marble staircase, but the earthquake had collapsed the staircase, leaving only the iron banisters intact. With an unbelievable amount of courage and no second thoughts, Eastwood scaled the wrecked staircase, holding onto the iron railing and putting her feet between the rungs.

The previous year, concerned about the possibility of a fire, she had selected type specimens of the academy's most prized plants and separated them from others in the collection. She then placed these type specimens in a specially built wooden box with glass doors that she would be able to lower, with a rope-and-pulley system, through an outside window to the ground below. But the rope-and-pulley system had not yet been constructed. When she finally reached the sixth floor and saw the special collection, she found the case had been knocked over and the boxes containing the prized plants scattered across the floor. She would have to improvise.

She found two large aprons and laced them together using rope and string—she routinely saved rope and string—then made a line sufficiently long to reach the ground floor. With the help of another academy worker, she repeatedly lowered and raised the aprons, each time sending down another bundle of valuable plants, until every specimen in the type collection was safely on the ground. She was back on the ground floor when she remembered she had left out a key plant—a saxifrage, a smallish plant that grows close to the ground and is found in alpine areas—of which she was preparing samples to send to the New York Botanical Garden. So she again climbed the iron banister and retrieved the saxifrage. She took the entire collection outside, and after a long argument with soldiers who were preventing people from crossing back and forth across Market Street—flames were meanwhile literally starting to engulf the academy building as the fires traveled across the city at increasingly alarming speeds—she found a man with a wagon and was able to take her plants back to where she lived and stored them, at least temporarily, in the hall of her house.

Eastwood spent the night alone in her loft apartment, unable to sleep. The fire front had stalled about a half-dozen blocks away, though the flames, as she recalled, were still bright enough to read a book by the red glow. Several times during the night her apartment rocked as firefighters dynamited nearby buildings, hoping to form a firebreak. She made use of her time by organizing her most valuable possessions in case she had to make a quick departure.

She packed her best clothes and extra underwear, her mother's Bible, a family photo album, and her typewriter. She had two Navajo blankets that she planned to use to cover the plants if they should be exposed to outdoor night weather. She also packed her revolver in case she had to sleep outside.

By morning, the fire was again advancing. By midday, it had swept through Chinatown and had consumed the mansions at the top of Nob Hill. Unable to stay any longer at her place, Eastwood, with the help of two young men she knew, found another wagon and hauled her plants to the Army post at Fort Mason on the north edge of the city next to the bay, figuring, as many in the city had, that this would be the last place to stand against the fires.

After that, she tried to return home but was stopped by soldiers. Fortunately, she had taken her packed belongings and, finding other people she knew, hired a local fishing boat to take them across the bay to Oakland. Eastwood spent the night sleeping in the attic of a friend. Several other people who also knew the homeowner joined her. Here it is tempting to imagine that somehow she and Gilbert met up that night—and they might have, though there are no letters or diary entries to support it. (They had always been discreet in their affair.)

All that is known for sure is that except for the few possessions Eastwood took from her apartment the previous night, she lost everything else—all her furniture, her books, and many treasured items, not to mention the countless plants she was unable to save from the academy before it burned. The only personal item she managed to save from her office was a Zeiss hand lens that she always wore as a necklace when she went to work.

And, of course, she saved a rare collection of plants—1,491 specimens, to be exact—which would become the foundation of the academy's new botanical garden.

Lawson was able to get a message to the governor soon after the governor's arrival in Oakland at 3:00 A.M. on the morning after the earthquake to oversee response to the disaster—the governor's arrival had been delayed many hours due to confusion along the railroad lines. Following the details of Lawson's message, the governor quickly established a scientific commission to study the earthquake and its effects and signed a number of passes that allowed the bearers to travel anywhere, which included across police or military lines, to carry out such work. Gilbert received one of the passes.

And so, on Friday, April 20, Gilbert finally made it across the bay to San Francisco. It is known that he walked along Market Street, where the fire had already passed, and at some time during the day made it to Washington Street and may have seen Eastwood's apartment consumed by flames. He also went to Van Ness Avenue on the west side of the city, where the last attempt to stop the fire was being made.

"The houses opposite were blistered and had glass broken," he wrote in his notebook of the scene, "and at one place the fire broke across, to be checked at Franklin St."

The plan for fighting—and stopping—the fire was this: Soldiers had set up a line of cannons on the west side of the avenue and firefighters had laid dynamite charges at the bases of buildings on the other side. In one great fusillade, followed by a pattern of directed explosions, a two-mile-long section consisting mostly of elegant Victorian mansions was to be demolished. The hope was this would produce a sufficient firebreak to stop the fire.

Whether the firebreak worked is still debated because a few hours after the mansions were destroyed along Van Ness Avenue, during the dark morning hours of Saturday, April 21, the wind direction shifted and a light rain fell. No matter what the reason, by midday the fire was finally out.

In all, three-fourths of the city had been consumed by flames. An official report would list the number of buildings destroyed as 28,188. And over half of the city's population of nearly 400,000 had been displaced.

But the people of San Francisco persevered and immediately began to rebuild their city. And the members of the newly founded commission began the task of collecting and compiling information about the earthquake with the intention that earthquakes would not be ignored again.

By Tuesday, April 24, there was enough information for Lawson to hold the first meeting of the earthquake commission, which met at noon on the university campus in Berkeley. Present were Lawson, Gilbert, astronomer Leuschner, and another astronomer, Charles Burkhalter, director of Chabot Observatory in Oakland. More members would be added later.

They divided themselves into two committees. Leuschner and Burkhalter would collect all the information that could be used to determine the exact times the earthquake wave had passed through various places in California. Like the timing of the wave passage for the 1886 Charleston earthquake, most of these determinations would come by noting when pendulum clocks stopped and where. And then, using these determinations, Leuschner and Burkhalter might be able to compute the speed of the wave—as was done for the 1886 earthquake—and if possible track the wave's progress backwards and pinpoint where the earthquake had originated.

The other committee—Lawson and Gilbert, but since Lawson was chairman of the commission, most of the initial work of this second committee would be done by Gilbert—would search the countryside and talk to local people and determine what surface changes had occurred as a direct result of the earthquake. Gilbert would seek out where cracks had formed, describe the type of damage done to buildings, dams, and roads, and identify where pipelines, water lines, and gas lines had been disrupted. It was recognized early that completing this work would require a small army, so many people were eventually enlisted, in particular a number of students at the university at Berkeley and at Stanford University.

At the time the commission first met, it was known that the most severe shaking was not confined to a small region—as Mallet had recorded after an earthquake in Italy in 1857 when damage was reported over an area of more than ten thousand square miles. What was surprising was that the greatest damage from this earthquake had occurred along a long line that ran for a long distance along the California coast and that the most severe destruction had not been in San Francisco, but in the communities immediately north of the Golden Gate. So that is where Gilbert started his investigation. And on at least one of these trips, he was accompanied by Alice Eastwood.

Her role was much more than that of companion—or to satisfy her own need to understand the earthquake that had nearly destroyed much of her life's work. She was familiar with the region and Gilbert was not. She had spent many a long weekend when she first moved to California walking the roads and searching the forests and the expansive grasslands for specimens to add to the academy's botanical collection, making her a little-recognized asset and integral figure in the early years of earthquake science.

She was staying at a friend's house close to the university when Gilbert arrived one morning driving a one-horse buggy. They rode to the Berkeley train station, where they took a scheduled train to Sausalito. At Sausalito, they hired a horse and wagon and drove over a low range of mountains to Point Reyes Station at the north end of Bear Valley, a straight and narrow trough that separates the triangular peninsula of Point Reyes from the rest of North America. It was here that they saw one of the first peculiarities created by this recent earthquake.

A train had been derailed. The 5:15 train for San Francisco was ready to depart on the morning of April 18 and the conductor had just swung himself onboard when the train lurched to the east. That was followed by an even greater lean to the west, so much so that the entire train—a steam locomotive and two passenger cars—was laid on its side without a single person being injured and without a single glass window being broken.

Gilbert and Eastwood traveled the full length of Bear Valley. At Inverness, a fishing village just north of the valley, they saw a long pier that, originally straight, now had a tight curve to it. Questioning workmen who were repairing the pier, they were told that the ruined segment of the pier had telescoped together so that the pier was now twelve feet shorter than before the earthquake.

At the southern end of Bear Valley, at Bolinas, they found a line of eucalyptus trees—set up to mark a property line—that now had an abrupt offset of ten feet. Similar offsets were seen elsewhere, along fence lines and across roads. And the direction of the offset was always the same—no matter which side of a fence line or a road one stood, the other side had moved to the right.*

They stopped at the dairy ranch of longtime resident Walter Skinner, also located within Bear Valley. Here they found an astounding number of offsets. A row of raspberry bushes in a garden was offset 14½. A path in the garden that Skinner and

* To understand why everything seemed to move to the right, consider this: Imagine you and a friend are on opposite sides of the San Andreas Fault and you ask your friend to stand directly in front of you and to face you. Now ask your friend to move to *your* right. Obviously, from your viewpoint, your friend has shifted to the right. More important, from your friend's viewpoint on the other side of the fault, *you* seem to have moved to the right.

his wife attested had been opposite the front door of their house before the earthquake was now 15 feet out of place. A fence bounding a cow pen had moved 15½ feet. But the strangest feature was a barn that had broken free of its joists and foundation and moved an incredible 16 feet. And running directly under the barn, as well as between the posts where the fence that formed the cow pen was offset, between the ranch house and its now displaced path, and along the exact line where the rows of raspberry bushes were displaced, was a curious furrow-like ridge that stood about a foot high and was about 3 feet wide, where the ground surface had been torn up and heaved and that looked, for lack of a better comparison, like a giant mole had dug a straight line.

In fact, this mole-track ridge could be followed for miles. Gilbert and Eastwood had first seen it where the eucalyptus trees at Bolinas were offset. They could trace it through the points where fence lines and roads were offset near and through Skinner's farm. The ridge ran nearly continuously the entire length of Bear Valley, a distance of 15 miles, from the marshy land at the head of Tomales Bay in the north to the sandy spit at Bolinas Lagoon in the south. And, as Gilbert noted, the ridge was "remarkably straight."

He was traveling that day with his camera and took what soon became several iconic photographs of the effects of the 1906 earthquake. One of these photographs shows the train lying on its side at Point Reyes Station, a man and a young girl looking off in different directions while a white dog seems disinterested in the whole affair. Gilbert also took several photographs of the mole track. One shows Alice Eastwood standing next to it—the mole track is running straight, the sod is overturned—on the slope of a small grass-covered hill.

It is one of the few photographs of her for which we know approximately when the photograph was taken, in late April or in May of 1906. And so it is worth pausing and considering how she looks. She is wearing a hat decorated with flowers, one of her normal habits, disdaining the current fashion of large bird plumage on one's hat. She is dressed in a light-colored shirtwaist and a dark ankle-length skirt, both in style at the time. Overall, there is a frumpish look about her because there seems to be a slightly bulging midsection. Is this the same woman who, just weeks before, *twice* scaled a broken staircase to the sixth floor of the academy building? Who, years earlier, had traversed 40 miles or more a day in forests and brush land searching for rare plant specimens?

Yes, it is.

A little investigation shows that Eastwood was an efficient traveler. When she went on a short trip, such as this one, she carried her nightgown as a roll around her middle, accounting for her somewhat stout appearance in the photograph—a fitting image of someone with a practical and passionate mind who had recently saved more than a thousand irreplaceable plants and had forgone the rescuing of her own possessions.

Gilbert sent preliminary reports of his investigations to Lawson at Berkeley, as did John Branner, head of the geology department at Stanford University, who sent out a small army of students to work much of central California.

Lawson was struck by the sameness of the reports. Dislocated fences, roads, railway bridges, tunnels, dams, and pipes were found from Point Arena on the coast north of San Francisco to the Spanish mission at San Juan Bautista east of Monterey Bay, a distance of 190 miles. Moreover, every dislocation was in the same direction—an offset to the right—though the amounts of offset varied from a maximum of 20 feet near the Skinner ranch in Bear Valley to lesser amounts to the north and the south.

And because buildings in the remote community of Petrolia, sited in the redwoods a few miles south of Eureka in Humboldt

County and lying along a projection of the rupture northward from Point Arena, were severely damaged, Lawson suggested the total length of the rupture was probably 270 miles—a speculation that would be confirmed decades later.

To put this in perspective, though dozens of earthquakes since 1906 have released more energy and have caused more seismic damage than the San Francisco earthquake, none have produced a longer line of surface rupture. In 1939, an earthquake in eastern Turkey approached that distance when the ground broke for almost 200 miles. Four years later, again in Turkey, another broke the ground a similar distance. Not until 2001 did an earthquake produce a surface rupture comparable in length to that in 1906: the Kokoxili earthquake along the Kunlun Fault in western China and northern Tibet, which created a line of surface rupture 240 miles long.

Lawson also took particular pleasure in noting the location of the rupture. Gilbert had traced it as a straight line through Bear Valley as far south as the sandy spit at Bolinas Lagoon, where it disappeared under the sea. Others found that it reappeared on land near Mussel Rock. From there, the rupture ran for several miles, as a mole track, through the line of small ponds found by Palache and Lawson a decade earlier and continued along the San Andreas Valley. And so the conclusion was inescapable: The fault Lawson had discovered in 1892 and declared active had slipped and slid during the 1906 earthquake.

Moreover, he now realized that what had ruptured in 1906 was only a segment of a much longer feature, one that he and his former student Fairbanks had followed across southern California just months before the earthquake. As Lawson took out his maps, he must have been amazed: Cutting obliquely across the entire length of the Coast Ranges and continuing across southern California from Humboldt County to the Salton Basin in the Colorado Desert, a distance of more than 800 miles, was a single geologic feature—the longest one known at the time—the San Andreas Fault.

Two months after the earthquake, the other committee, comprised of the two astronomers Leuschner and Burkhalter and charged with compiling and analyzing all information related to the time of passage of the earthquake wave across California, had made little progress, so Lawson gave their task to someone else who was added to the commission: Harry Fielding Reid of Johns Hopkins University in Baltimore.*

At first glance, Reid seemed to be an odd choice. He was from the East Coast and as far as can be determined had never been to California. He was an expert on glaciers, having done fieldwork in the Swiss Alps and in Alaska, but did have a minor association with earthquakes. Since 1902 he had been hired by the United States Weather Bureau to examine seismographs collected at a handful of weather stations, of which one was at Johns Hopkins. For that work, to cover his salary and his expenses, Reid was paid $100 a year, an indication of how minor this work was perceived to be.

Reid arrived in California in late June and began to sift through the scores of timing reports, mainly of pendulum clocks stopped by earthquake shaking, finding that the vast majority were known to an accuracy of a minute or so. Because of the great speed of the wave, in order to calculate the earthquake's place of origin, Reid knew he needed the time of passage to be within one or two seconds in order to compute where the most violent shaking had originated. By that requirement, only four records qualified: one maintained by the time inspector of the North Shore Railroad in San Rafael,

* Reid was the eighth and last member added to the earthquake commission. In addition to Lawson, Gilbert, Leuschner, and Burkhalter, the other members were geologist John Branner of Stanford; the director of Lick Observatory, William Campbell; and George Davidson, the most august member of the commission—aged 81 at the time of the earthquake—who had conducted many of the original land surveys along the Pacific coast from Alaska to Panama and who was a past president of the California Academy of Sciences.

located 20 miles north of San Francisco, and those of the standard clock used at the Navy Yard at Mare Island to set ship chronometers, located near the north edge of the bay, and two clocks at astronomical observatories, one at the Students' Observatory on the Berkeley campus and the other a small clock that had sat on the director's desk at Lick Observatory.

From those four time records, Reid determined that the most violent shock had originated beneath the town of Olema near the Skinner ranch, a place, he noted, where, according to eyewitness accounts, "the violence of the shock was probably as great as anywhere," and where investigators, primarily Gilbert, had reported the largest amount of surface offset. For years Reid's determination was accepted as the earthquake's origin—its epicenter—and today, people who live in Olema take a measure of pride in announcing that they live where the 1906 disaster began. But they are wrong.

Ninety-six seismographs around the world recorded the passage of the earthquake wave on April 18, 1906. The most distant station was on the island of Mauritius in the Indian Ocean, a direct-line distance through the earth from San Francisco of 7,827 miles. It took 84 minutes for the first disturbance to reach this station. That was followed by more than three hours of vibrations, many appearing as distinct wave trains.

Reid knew of these records, but he did not know how to analyze them; he did not know how to identify a specific wave train on one record with the same wave train on others. Today, we do. A sophisticated analysis of the wave trains recorded on these distant seismographs shows that Reid and his four time records gave an epicenter that was 20 miles too far north of the actual one.

The main shock in 1906 originated beneath the sea 2 miles west of the shoreline at Golden Gate Park—8 miles west of downtown San Francisco—at a depth of 6 miles. Furthermore, from the individual scratches and squiggles traced out on the seismographs, which may seem incomprehensible to the layperson but are decipherable by a specialist, it is now possible to show that once the earthquake began, the rupture propagated both north and south at a speed of about 2 miles per second, about ten times the cruising speed of a commercial jetliner—though for some unknown reason that is still debated, the rupture grew slightly faster to the north than to the south—reaching the far ends at Petrolia and at San Juan Bautista in about one minute.

In the weeks immediately after the earthquake, some scientists, notably Branner at Stanford, voiced the standard explanation that the earthquake had been caused either by a local heating and expansion of the crust by rising hot rock or by a slow cooling and contraction of the crust to produce a strain in the other direction. But those opinions soon changed after it was realized how low the rupture was and how consistent the horizontal slip. By late May, Branner was telling audiences that it was his belief "that the recent shock was caused by the slipping of an old fault." Seldom has a disaster produced such a dramatic shift in scientific opinion, but after 1906, most scientists accepted Gilbert's suggestion made in 1884—and restated in 1893 by Bunjiro Koto after the Mino-Owari earthquake in central Japan—that the rupture had been the *cause,* not the *effect,* of the earthquake. But that introduced a new question: If neither the earth's internal heat nor its slow and steady cooling and contraction was the source of seismic energy, then what was? Reid soon provided the answer.

Because of the early economic importance of San Francisco Bay, accurate maps of the region were essential. They were prepared twice, first in the 1860s and again in the 1890s. Both times, they were based on determining, through land surveys, the precise locations of a network of control points, most sited on hilltops. Because of the dramatic shift in ground positions recorded by Gilbert and others along the 1906 rupture, a third survey was conducted. It was completed in the spring of 1907.

Reid compared the position of all points surveyed in the 1890s and again in 1907. He found that every point on the west side of the San Andreas Fault had moved north with respect to those on the east side of the fault. Furthermore, the amount of movement decreased with distance from the fault, so that if a point a mile west of the fault had moved 11 feet to the north, one 4 miles west had moved 8 feet, and on the Farallon Islands, 23 miles west of the rupture, a point had moved northward nearly 6 feet.

Then came the surprise.

When Reid compared how points had moved between the 1860s and the 1890s, he found that most points showed little or no movement *except* for the westernmost point on the Farallon Islands—which had moved a remarkable 5 feet north *without* an earthquake. Reid knew immediately what this meant.

For an unknown reason—which would not be explained for another 60 years—the Farallon Islands, situated out in the Pacific Ocean more than 20 miles *west* of the San Andreas Fault, were constantly moving north at about two inches a year with respect to other points measured around San Francisco Bay, but without earthquakes. That slow and steady movement set up a strain of captured energy that continued to build and build until it exceeded the finite strength of the rock. Then, in Reid's words, there is a "sudden fling" as the rock fractures—along the San Andreas Fault—and the opposite sides of the fault slide to new positions of no strain, releasing the pent-up energy. It was, again according to Reid, similar to winding an elastic spring inside a watch. As the spring is wound tighter and tighter, elastic energy builds until the spring breaks. The same happens inside the Earth. As the Farallon Islands and their surroundings continue to move north, elastic energy builds until it is released as an earthquake. Then the process starts again.

Reid called his idea "the elastic rebound theory of earthquakes," and it has been at the heart of understanding—and attempts to predict—earthquakes for more than a century. A simple calculation shows that if the Farallon Islands are moving north at a rate of 2 inches a year, then after a century the islands are about 16 feet farther north—the same as the amount of horizontal slip at

Olema during the 1906 earthquake. And so—and this was a big step forward—major earthquakes should repeat along the San Andreas on average every hundred years or so. Furthermore, Reid realized, as Lawson and Gilbert and others did, the buildup to an earthquake could now be recorded by placing a line of survey points across the fault and measuring how they moved. When the strain came close to the breaking strength of rock, an earthquake was imminent.

In practice—as an additional century of scientific studies have shown—earthquakes are much more complicated than this. But this was a start—a crucial start. And so it is not a stretch to claim that our basic knowledge of earthquakes—what they are, the source of the energy they release, how they might be predicted—comes from the remarkably long and remarkably straight 1906 rupture of the San Andreas Fault.

Two massive scientific reports were prepared for the 1906 earthquake, the first published in 1908 by Lawson and the second in 1910 by Reid. These tomes set the standard for future studies of earthquakes. Eyewitness accounts are abundant. The type and location of seismic destruction is described in detail and illustrated by hundreds of photographs. There are detailed descriptions of the faulting and reports on shaking intensity. The ground rupture is also described, as are the results of the land surveys conducted before and immediately after the earthquake. There are brief reports from the 96 stations where the earthquake waves were recorded on seismographs. Of special value, provided as an atlas, are tracings of the waves from the 96 seismographs and 40 detailed maps that show exactly where the ground rupture was found.

In 1909, Lawson was elected the second president of the Seismological Society of America. The society officially organized on August 30, 1906, while aftershocks were still rocking the region.

The founding membership numbered almost 300; today there are more than 2,000 members. In 1911, he taught the first university course in seismology in the United States.

Reid's life took an unusual turn in 1922 when a close friend, Edith Hamilton—the famed classicist and headmistress of Bryn Mawr School for Girls in Baltimore—took up residence in the Reid household. Hamilton was having a dispute with the school's benefactor, M. Carey Thomas, and after a shouting match with Thomas at Reid's house, Hamilton resigned as headmistress.

The next summer, Hamilton and Reid bought a summer house in Maine, where Hamilton and Reid's eldest daughter, Doris Fielding Reid, resided. Their relationship was described, according to the vernacular of the time, as a "Boston marriage," which meant two women were living together without the financial support of a man. Doris Reid took work as an investment banker, in which she succeeded, and Hamilton began writing her famous series of books on mythology. The first book, *The Greek Way*, was dedicated to Doris Fielding Reid. Both women were at Harry Reid's bedside when he died in 1944. One wonders whether any discussions with Reid about geology—and about his experiences in San Francisco after the earthquake and his groundbreaking (no pun intended) conclusion on elastic wave theory—crept into the mythological imagery used by Hamilton in her writings.

Alice Eastwood worked as a volunteer at the University of California for a year after the earthquake. She then set off traveling, using for financial support a small amount of money she was earning from a hotel in Colorado in which she had invested before she moved to California. For a while she worked, again as a volunteer, at the Smithsonian Institution, the New York Botanical Garden, and an arboretum near Boston. She also found time, and money, to visit botanical collections in London and Paris. In 1912, she was called back to California to resume her position at the newly rebuilt California Academy of Sciences, working at the new botanical garden in Golden Gate Park.

In March 1909, Grove Karl Gilbert developed a serious illness. It was diagnosed as apoplexy—a cerebral hemorrhage or stroke. "While my physician's tone is optimistic," he wrote to a friend, "my own impression is that my general physical condition is lowered."

But he did recover, slowly. Not until 1918 was he ready to declare his convalescence over, and when he did, he had something else to announce. In a letter to one of his sons, he finally confirmed what many had gossiped about. "Alice and I have been lovers for years."

He continued: "But for a long time I would not propose marriage because it seemed like asking her to give up a life that satisfied her to become the nurse of my broken health." Here he was probably thinking of the difficult 18 years he had cared for his late wife during her prolonged illness. But now, "my general health has so far improved that I am less ashamed to impose myself on Alice." And so he wrote to her and asked if she would marry him. She replied with a one-word answer: "Yes."

He left Washington, D.C., where he was then residing, and traveled to San Francisco to reunite with his new fiancée, stopping as he often did on a transcontinental trip for a short stay with his sister who lived in Jackson, Michigan. Here he suffered a second stroke. This time, he did not recover. He died on May 1, 1918.

Eastwood lived another 35 years, continuing her work as curator of botany. By the time of her death in 1953, at age 94, she had named 125 species of California plants. In her honor, her colleagues had named eight plant species for her, including *Salix eastwoodiae*, a willow found in central California's alpine regions, and *Eastwoodia elegans*, a yellow aster.

But to those who study seismology and are familiar with the 1906 earthquake, she is the woman, often unidentified, in a photograph standing next to the "mole track" north of Olema, California.

CHAPTER 4

BRIDGING "THE GOLDEN GATE"

I was up on the tower when the earthquake hit.
It was so limber the tower swayed sixteen feet each way.
—Alfred "Frenchy" Gales,
on the Golden Gate Bridge, 1934

After the 1906 disaster, Lawson decided he would build a house that could withstand both earthquakes and fires. He consulted a study made of over 1,000 houses that were located in communities south of San Francisco and within four miles of the San Andreas Fault to determine what type of structure had suffered the least damage. By far, brick and stone buildings had faired the worst. Each one, according to the study, had either collapsed walls or huge cracks, and in some cases were totally demolished. The houses that had survived the best were single-storied wooden structures, though these, of course, were the most susceptible to fire. So Lawson decided he would risk a new type of construction—and build a new house of reinforced concrete.

Concrete had been used by the ancient Romans to construct many different types of structures, but after the empire passed, its use was limited until the 1870s when French engineer Francois Hennebique pioneered the idea of wrapping concrete columns and of embedding concrete slabs with steel bars to give individual architectural elements a high amount of tensional strength. Hennebique's idea was quickly adopted elsewhere, most notably in New York and Chicago, but it was regarded with suspicion in San Francisco, where leaders of brick-layer unions, concerned about the employment of their members, dissuaded architects and building contractors from using it.

That changed after 1906, when it was clear that the few buildings in San Francisco constructed of reinforced concrete had performed remarkably well during the earthquake and the subsequent fire. Though the city still had limits on how high a structure could be built using the new technique, by June 1907, there were 78 reinforced concrete buildings under construction in San Francisco. As far as can be determined, Lawson was the first to use it to build a house.

He consulted with his friend and Berkeley neighbor Bernard Maybeck on how the house should be designed and how it should be constructed. Maybeck had joined the university faculty as a drawing instructor in civil engineering in 1894. Four years later, showing a remarkable talent in his drawings and originality in design, he was named the university's first professor of architecture. In 1896, while still an instructor, Maybeck had designed a Gothic-style house for Lawson that was built south of the university and where Lawson and his family had been at the time of the earthquake, the only damage being a collapsed brick chimney.

Maybeck's career eventually soared. During his long career, which covered more than 60 years, he would be the chief architect on more than a hundred major projects in the San Francisco area. His most famous design is the Palace of Fine Arts constructed for the 1915 Panama-Pacific International Exposition. This giant rotunda—which during the 1930s housed 18 lighted tennis courts—is one of only a few surviving structures of the Exposition and the only one still situated on its original site. In 2005 it was added to the National Register of Historic Landmarks, and in 2009 a seismic retrofit was completed to insure—it is hoped—that this iconic architectural feature of San Francisco will survive all future seismic shaking.

Inspired by the idea of combining the Mediterranean-style setting and climate of Berkeley with the fact that Mount Vesuvius had erupted just two weeks before the San Francisco earthquake, Lawson and Maybeck settled on a design for the house modeled on a Pompeiian villa. They kept the imposing cubistic form and large rooms but abandoned the interior mural paintings, replacing them with multicolored diamond patterns scratched into the concrete walls. Instead of a tile roof, they had the building covered with a single large concrete slab angled at a low pitch. Five bedrooms occupied the upstairs floor, each room communicating directly with two sleeping porches.

The house was constructed on a tract of land called La Loma Park, a wooded area that Lawson and Maybeck had purchased in 1900 and that is located immediately north of the university. This area did not yet have paved roads, so mule-drawn wagons were used to haul the concrete and steel bars and other material to the site. The concrete was mixed by hand in batches of one cubic yard and carried in buckets by a small army of local laborers. The final structure has outer walls that are three feet thick.

Maybeck covered the outside of the building with stucco that he had colored a light red on the lower part and buff on the upper so that it resembled a Pompeiian villa. He also had muddy water thrown onto freshly painted walls to give the house an appearance of age.

The building's resistance to fire was tested decades later on September 17, 1923, when a grass fire in nearby Wildcat Canyon spread into the Berkeley city limits. Lawson was away, attending a college reunion in Toronto. Well-intentioned neighbors broke into his house and removed much of the furniture and many valuable objects. But the progress of the fire was too fast, and the would-be rescuers had to abandon Lawson's furniture and other valuables on the street, where they were consumed by flames.

In all, the fire burned through 50 city blocks and destroyed nearly 600 homes. But the Lawson house withstood the flames. Everything inside the house, including his library—he was a collector of rare books—and his personal papers, escaped unharmed. Fortunately for the scientific and bibliophilic communities, the neighbors had only tried to "save" the furniture by removing it from the house.

The ability of Lawson's house to withstand strong earthquake shaking has yet to be tested—but it will be. Lawson located his house on a hillside with a grand view of the bay, knowing it was within 200 feet of the active Hayward Fault.

As the scientific investigation of the 1906 earthquake was progressing—and while he was building his house, which was completed in 1909—Lawson also investigated the effects of other recent earthquakes, notably the 1868 event that had produced the strongest shaking and the most damage prior to 1906. In talking to people who had experienced that event, he was able to uncover the fact that a rupture had formed at intervals along a nearly straight line from the east side of Mills College in Oakland southward through the center of the town of Hayward (which was then known as Haywards), where damage had been the most severe, to the small isolated community of Warm Springs, which today is the home of many high-technology companies and is part of the incorporated city of Fremont.

From his inquiries, Lawson came to realize that if the trace of the 1868 rupture was projected north, it would pass through a long narrow valley, one that was similar in length and in width to the narrow valley that contains Olema and Point Reyes Station and that the 1906 rupture ran through.* And if the trend of the valley was extended north, it would pass along the western base of the Berkeley Hills.

In all, Lawson realized that a second active fault—the Hayward Fault, as he called it—ran along the entire length of the east side of San Francisco Bay. As to the risk of building a house right next to an active fault, even after the 1906 disaster, Lawson shared the opinion of many others: The seismic damage caused to buildings in 1906—as well as in 1868—had been the result of poor construction. If a house was built, in Lawson's words, of "honest construction," then any house could be made "practically earthquake-proof." And that idea could be extended, so he argued, to any other type of structure—even one that was built near an active fault.

As a case in point, one of the most imposing structures constructed during the two decades that followed the 1906 earthquake was a football stadium at the University of California, dedicated as California Memorial Stadium. Even before its completion in 1923, it was known that the stadium straddled the Hayward Fault. So the architects, of a practical bent, designed the stadium as two halves that would slide apart in the event of a major earthquake.

But Lawson's house—which was, essentially, a concrete bunker—and California Memorial Stadium—where specially designed joists were used—were exceptions. For the most part, San Francisco was rebuilt after the 1906 earthquake without consideration of what a future earthquake might do to the city because, as seemed evident—and was certainly pragmatically true—the city had been destroyed primarily by fire, not by earthquake shaking.

So new fire codes were written and approved, but regulations that required the construction of earthquake-resistant buildings were not, because it was generally agreed that such

* The California Department of Transportation has taken advantage of this natural corridor and has rerouted the Warren Freeway, officially California State Highway 13, which is used by thousands of commuters every day.

requirements were already covered by regulations that buildings withstand wind gusts.

This was the general thinking behind construction in San Francisco at the time—until it came to the most challenging construction project of the era in California. In that case, concern about the San Andreas was of paramount importance because the project involved the building of a great bridge to span the mile-wide entrance to San Francisco Bay known as the Golden Gate.

The narrow strait that connects San Francisco Bay to the Pacific Ocean was once a river gorge that formed during a low stand of the sea during a recent ice age—the ocean shoreline then far to the west. During those frigid times, a large cap of ice covered much of North America and Europe. The subsequent melting of the ice cap caused the level of the sea to rise, drowning what, along the central California coast, were coastal hills whose tops are now the Farallon Islands. It also drowned the river gorge, giving us what is one of the most picturesque land and seascapes in the world. Eventually, the drowning of the gorge produced something else, something uniquely human: a temptation and desire to span the strait with a bridge.

In 1921, Joseph Strauss, a prominent Chicago engineer and noted bridge builder, printed and distributed, at his own expense, a brochure entitled *Bridging "the Golden Gate."* In it, he described how such a bridge could be built. Years followed as he revised his designs and lobbied for the necessary local, state, and federal approvals. Finally, on December 4, 1928, the state of California established the Golden Gate Bridge and Highway District, which would oversee the building of such a bridge. Less than a year later, on August 15, 1929, the board of directors of the highway district officially named Strauss as its chief engineer.

Early on, the state of California realized that the bridge would have to be financed by private money by the selling of public bonds. Thirty-five million dollars would be needed. But to issue such bonds, the state required the approval of voters. And it was Strauss who led the charge to get such approval.

With money supplied by the highway district, he bought newspaper advertisements and billboard space to promote the bridge. He convinced local politicians to make speeches in favor of the bridge. Newspaper publishers were paid to print editorials in support of the bridge. Strauss hired experts to produce massive reports that attested to the bridge's utility in expanding the region's economy and in providing people with more mobility, now that private automobiles were popular. Eventually, a key question was asked: Would the proposed bridge, which would be the longest single span ever constructed, survive a major earthquake?

Less than a generation had passed since the 1906 earthquake, so memories of that event were still in people's minds. But Strauss knew there was a highly regarded local authority who had the expertise to prepare a geology report and that the public would accept immediately what he said. So Strauss hired Andrew Lawson as his geology consultant.

The Berkeley professor set upon doing the work in his usual diligent and thorough style and presented the report to Strauss on February 12, 1930. The chief engineer exploded when he read it.

In the report, Lawson described the rocks exposed on the opposite sides of the Golden Gate. At Lime Point, on the north side, was dense basalt that would provide a stable and strong platform for a bridge pier. But on the south side, at Fort Point, the ground was covered by a soft, easily eroded, slippery green rock known as serpentine.*

No excavation or deep drilling had yet been done, so Lawson did not know how far downward the friable green rock extended. But

* Technically, the rock is called *serpentinite* and is comprised of a group of minerals known collectively as *serpentine,* but the word "serpentine" is so widely used to refer to the rock outside scientific circles that I have used it here.

he was concerned about the stability of any massive structure built on this site. In his opinion, he wrote in the report, the south pier of the bridge would "have to be designed to depend upon the dead load rather than upon the tensile strength of the rock." He continued by reminding that "once or twice a century, it may reasonably be assumed the region of San Francisco Bay will be shaken by a violent earthquake," and added, again as a crucial reminder, that the trace of the San Andreas Fault, "upon which [there was] a sudden slip in 1906, with disastrous results to the City of San Francisco," was just a few miles west of the proposed bridge.

Those statements were enough to put into question whether the construction of a bridge was technically feasible—and whether Strauss should receive the 35 million dollars to build it—but it was what Lawson wrote next that worried Strauss more.

Lawson confined himself not only to a consideration of the geology of the site but also to the severe weather conditions. The bridge would be exposed to salt-laden air and battered by strong winds and the occasional Pacific storm. Such conditions, in his opinion, meant the bridge "would have to be replaced once or twice a century owing to deterioration by rust."

That last statement—in which Lawson had clearly exceeded his expertise and had gone beyond what Strauss had requested of him was enough for the chief engineer to refuse the entire geology report. The next day, March 7, 1930, Lawson announced that he was resigning as consulting geologist on the project to build the Golden Gate Bridge.

Strauss was not a man of charm or diplomacy, but he knew that an unfavorable geology report was better than no geology report at all. So on March 8, the day after Lawson resigned, Strauss sent one of his assistant engineers, Charles Derleth, who was also chairman of the Department of Engineering at Berkeley, to talk with the geology professor and get him "to modify his position."

Derleth and Lawson met for two hours. Lawson refused to change the report. Desperate to keep Lawson on as consulting geology, Derleth offered a compromise. The report would be accepted in its entirety *if* Lawson agreed to add a sentence at the end that read: "Any earthquake so violent that it would destroy the bridge would also destroy San Francisco."

Lawson agreed. And the report was published.

On November 4, 1930, after a half-million-dollar campaign blitz was waged by Strauss in favor of the bridge, voters gave their approval by a margin of three to one to have public bonds issued to finance the building of the Golden Gate Bridge. Twenty-eight months later, on Sunday, February 26, 1933, at a ground-breaking ceremony on the grounds of the Presidio, Strauss was given a spade and turned over the first patch of earth that officially began construction of the bridge.

Between the time Lawson was initially hired by Strauss to be consulting geologist and the day of the ground-breaking ceremony, his life changed in a dramatic way. It began with the death of his wife while he was preparing his geology report.

Ludovika von Jansch, the daughter of a judge, was born in 1864 in Brunn, Moravia, in what is today the Czech Republic but was then part of the Austro-Hungarian Empire. She immigrated to Ottawa, Canada, and it was there that she and Lawson met and married. Ludovika never learned to speak English and conversed with her husband only in German. Lawson's sister Katherine, three years his junior and the same age as Ludovika, considered her sister-in-law to be "a nervous creature." No one in the Lawson family ever got personally close to her. After nearly 20 years of marriage and the raising of four children, she and her husband started arguing frequently. Ludovika finally moved out and left Berkeley and found a house in Carmel. Lawson remained at his job, continuing an intense schedule of teaching, research, and travel. In 1929, Ludovika became seriously ill and, after a prolonged hospitalization back in Berkeley, she died on Christmas Day.

Four months after her death—and after completing his geology report and having it reluctantly accepted by Strauss—Lawson left California for several months. At age 70, he still presented the aspect of a spry and energetic man. His friends commented on his ability to maintain a slender figure and a muscular tone. His hair was now white, though thick with only a little bald on top, and his eyebrows bushy. He wore a wide walrus mustache, having recently cut off his beard, which he would not let grow again for another ten years. On this trip, he first went to Morocco, where he stayed for several months working with two French geologists who were studying the Atlas Mountains. After that, he stopped in Canada to attend a scientific meeting and to see a lifelong friend, William Collins, director of the Geological Survey of Canada.

Unknown to Lawson as he prepared to return to Berkeley, his friends had decided that they would shower such affection and warmth on him when he returned that he would forget about his deceased wife. But *he* had a surprise for *them*. When he returned to Berkeley, he announced that he had remarried.

She was Isabel Collins, William Collins's daughter. She and Lawson had first met three years earlier on a geological expedition in Canada led by her father and for which she had been the expedition's assistant cook. It was her first time on such a trip and Lawson had taken the time to instruct her on some of the pleasures of such journeys, such as how to handle a canoe. On his stop in Canada during his return to Berkeley from Morocco, they resumed their acquaintance and an intense romance developed. They secretly married; she informed her parents by telephone after they left Canada and were driving to California. Isabel Collins was then in her last year of college at McGill University in Montreal, where her major course of study was home economics. She was 21 years old.

Once they arrived in Berkeley, their May-December romance and marriage became a frequent topic of gossip, whispered about by neighbors and occasionally finding its way into the pages of national newspapers. The new bride moved into Lawson's earthquake-proof and fireproof house near the university, but she thought the thick walls and the limited lighting were depressing, so her husband had a new house of standard design and construction built and that was where they resided. But eventually she tired of Berkeley, so a third house was bought in Carmel, and that was where Isabel lived during the summer months, swimming and writing long letters to her husband, whom she affectionately called "Skipper," while he responded to her letters by sending her poetry.

It was during this time, during the early years of their marriage, the question arose: Given its proximity to the San Andreas Fault and the unstable character of the green serpentine, should the Golden Gate Bridge be constructed?

For a variety of reasons, mainly economic and engineering, involving controlling costs and dealing with the reality of the finite strength of materials that would be used to build the bridge, the central span of the proposed bridge, which would be a single-suspension structure, could not exceed 4,200 feet. Simple arithmetic showed that in order to cross the mile-wide strait, one of the piers had to be built at least 1,000 feet from shore. Soundings showed that out from Lime Point on the north side, the water deepened quickly to more than 300 feet, so the north pier had to be built on land. But on the south side there is an underwater shelf that slopes slightly, so that at 1,000 feet from shore, the water depth is only 60 feet. It was there, at the edge of the shelf, where the south pier had to be constructed. That meant not only would the south pier rest on serpentine, which Lawson in his geology report had described as "in a rather badly sheared condition," but the foundation for the bridge had to be excavated to a yet-to-be-determined depth *underwater*!

The plan was to build a watertight, oval-shaped wall of concrete, 30 feet thick, large enough that it could enclose a football field, which would extend from the 60-foot depth of the sea floor north of Fort Point up to and extending 15 feet above the level of high tide. Then the seawater would be pumped out and—the interior of the oval structure being then free of water—workers would begin the difficult task of digging down, first through mud, then through serpentine, to a solid foundation. Lawson originally thought the workers would have to remove about 25 feet of serpentine to find competent rock, but his estimate proved to be nearly a factor of two too little.

It took almost a year to build the oval structure, to remove the seawater, and to excavate through *40* feet of serpentine. The foundation for the south pier was now at a depth of 100 feet below sea level. At that level, engineering tests showed that the serpentine should be able to support the hundreds of pounds per square inch of pressure the steel bridge would exert on the rock. Workers had started to use concrete to fill the massive hole when a new objection was raised—from a highly vocal source—about the building of the Golden Gate Bridge.

Bailey Willis, head of the geology department at Stanford University, certainly had the credentials to raise an objection. Unlike Strauss, who Willis claimed was an engineer who knew nothing of geology, and Lawson, who Willis said was a geologist who did not understand engineering, the Stanford professor could claim expertise in both fields. He had graduated with dual degrees in geology and civil engineering from Columbia University and had done practical work in both fields. Included in his long list of accomplishments was a year-long geological expedition he had led through northern China, and another year in Argentina, where he advised the government on the routing and construction of railroads in the Andes. In 1922, relying on the work of Lawson and many others, he had produced the first map that showed active faults in California, declaring that earthquakes along those features were "as certain as thunderstorms in New York in early summer."

And if one examined his map, which was published by the Seismological Society of America, of which both he and Lawson had once been president, there was an active fault that passed directly beneath where the south pier was under construction.

Willis first went public with his concern in late 1933, saying that the south pier, as it was then positioned, sat "on a 'pudding stone' of serpentine." Here the rock was so weak, Willis went on, that "even vibrations from the San Andreas Fault might cause the pier site and its 200,000-pound load to slide into the channel." Then there was the matter, as he also pointed out, of an active fault, a possible branch of the San Andreas, or so Willis suggested, running directly under the pier.

As if to illustrate his concern, soon after Willis had made these statements, trouble of a similar kind was being reported on the other side of San Francisco, where another massive bridge was under construction.

The Golden Gate Bridge and the Bay Bridge were being built simultaneously. But the Bay Bridge posed fewer engineering problems.

Its design consisted of two parts: One part ran from Oakland to Yerba Buena Island in the center of the bay, and the other part from Yerba Buena Island to a landing in San Francisco. Here tidal currents were slower than through the Golden Gate, and the impact of storms and other severe weather less intense. The floor of the bay was nearly flat and shallow. Concrete caissons were manufactured and sunk along a line that the bridge would run. Atop the concrete caissons, the steel piers would be built, upon which the weight of the bridge would rest.

But on January 25, 1934, a problem was reported at caisson W-6, the first one west of Yerba Buena Island. Surveys showed that the base of the caisson had slipped.

A diver was sent down. In the blackness of the bottom, the diver felt his way around the caisson, deciding that one edge was hung up on a huge boulder. He returned to the surface, dove again with a packet of dynamite, planted it where he thought it would do the most good, then resurfaced. The dynamite was exploded. Repeated surveys have shown the caisson has never shifted again.

But the incident at the Bay Bridge directed attention to the south pier of the Golden Gate Bridge. There were still opponents to completion of the bridge, notably from Southern Pacific Company, which owned the ferry lines that crossed the bay, and from lumber companies in northern California that saw their political power falling if the bridge was finished and they were no longer isolated from most of the rest of the state. So Willis's objections started to gain support.

Willis issued a letter of concern about the stability of the south pier, asking that the concrete that had already been poured be removed and that the foundation be dug down another 250 feet. Strauss asked Lawson for a response. In a private letter to Strauss, Lawson called the objections "pure buncombe," and in a public statement he charged Willis with being a "professional alarmist." He disputed Willis's claim that an active fault ran under the south pier. He said the green serpentine would be a firm foundation if workers excavated to a deep enough level. Willis responded by publishing a long letter, covering two full pages in the newspaper *The Argonaut*, that called for all work to halt at the bridge site. And then, as if on cue, two weeks after the letter appeared, the largest jolt to occur along the San Andreas Fault since 1906 struck near San Francisco at midday on October 2, 1934.

From the reports of the thousands of people who felt the earthquake, it was centered near Mussel Rock and consisted of two distinct shocks, separated by about ten minutes. Dishes and windows rattled in San Francisco. The only reported injury in the city was to a Miss Grace Williams, who was stepping down from a curb at Geary and Powell and fell and hurt her arm. A convention of 400 delegates of the American Federation of Labor, always a feisty crowd, was meeting in San Francisco at the time, and the delegates were in heated debates when the first shock hit. The convention hall suddenly went silent. After a minute or so, the delegates resumed shouting at each other—when the second shaking happened. During the ensuing silence, someone offered a motion to adjourn for lunch. It was passed by a unanimous vote.

The steel tower built over the north pier, the top 746 feet above sea level making the tower 200 feet taller than the Washington Monument, was almost completed when the earthquake struck, and a dozen or so men were working on the tower when it began to sway. One worker was Alfred "Frenchy" Gales, who had been a bus driver before becoming a bridge builder. According to him, during the swaying, the top of the tower moved 16 feet each way. He watched as "guys were lying on the deck, throwing up and everything." He decided, as the swaying continued, that if the tower failed he was going to ride the steel all the way to the water.

But the north tower did stand. Strauss, Lawson, and other engineers examined the site the next day, proclaiming the bridge had withstood its first earthquake test.

Still, Willis was not silent. He claimed the serpentine rock exposed at the south pier was only 100 or so feet in thickness, and beneath it was more friable sandstone. In the event of a repeat of what occurred in 1906, the weight of the bridge would cause the base of serpentine rock to shear off the lower sandstone and the whole bridge would collapse into the channel.

Strauss ordered workers on the south pier to continue to fill the oval structure with concrete. He also had several holes outside the oval structure drilled to a depth of 250 feet. Lawson examined the recovered cores, saying it was still serpentine and that it was dense and without fractures. But Willis persisted—and opposition to completing the bridge grew. To silence it, Strauss called on Lawson to stage a performance.

By then, 65 feet of concrete had been poured, though through the concrete eight inspection wells were still open that extended all the way down to the serpentine base. On December 7, 1934, Lawson made a descent down Well No. 3.

Before the descent he and Russell Cone, an inspecting engineer who would accompany him down, were bolted inside an air lock at the top of Well No. 3 while the air pressure was stepped up from normal outside atmosphere to 40 pounds per square inch on their bodies. That was to allow them to pass through an air lock partway down the well. Once the air pressure was at the proper level, the two men started down, climbing on ladders attached to the inside of the 4-foot-diameter well. There were electric lights all the way down.

At the bottom, more than 100 feet below sea level, they stood inside a dome-shaped room about the size of a typical living room. The glare of four 50-watt lamps lit the scene. The center of the floor was free of mud and rocky debris and stood about 2 feet higher than the edges. A puddle about 6 feet long and 2 feet wide was near one side of the room. Lawson knelt down and examined the rock. It was compact serpentine, remarkably free of any seams. He hit it with his hammer. As he would later report to Strauss, "It rings like steel." Then, in words that pleased the chief engineer, Lawson pronounced the serpentine foundation to be "all that could be desired." In his judgment, it "would support several times the pier load with perfect safety."

Lawson's report was given to the press. Bridge construction continued without much controversy, though there was one more seismic test.

On March 8, 1937, according to a press release from the university at Berkeley, an earthquake of "moderate intensity" was felt throughout the Bay Area. It was centered on the Hayward Fault. Sixty chimneys in El Cerrito, north of Berkeley, were damaged. People reported that dishes, glass bottles, and windows were broken elsewhere. It occurred at 2:32 in the morning, so no one was on the bridge. The next day an inspection was made and the bridge was declared by Strauss to be "untouched."

On May 27, 1937, the Golden Gate Bridge was dedicated and, for the day, open only to pedestrians. At noon the next day, President Franklin Roosevelt pressed a telegraph key in the White House and the bridge was officially open to traffic.

. .

Here it is instructive to examine the history of the reactions of the Golden Gate Bridge and the Bay Bridge to later seismic disturbances in light of their respective "births" at the same time that earthquake science were still in their early infancy.

On Friday, March 22, 1957, shortly before noon, an earthquake hit the Bay Area. It was centered near Daly City, possibly on the San Andreas Fault. According to the *San Francisco Chronicle*, in the city there was a "twisting, jarring side-rolling motion" that caused skyscrapers to sway visibly. The shaking was enough to cause people to run into the streets, some "sobbing hysterically." A person who was on the Golden Gate Bridge reported the bridge "undulated as in a fierce gale." But inspections showed no damage to it or to the bridge on the other side of the bay.

On February 9, 1971, an earthquake occurred north of Los Angeles in the San Fernando Valley. Never before had seismic shaking damaged a modern highway in California. Then, in just 12 seconds, nearly 70 bridges designed to resist earthquakes and thought by their designers to be quake proof were damaged. Seven bridges collapsed or were so badly damaged that they had to be replaced.

No shaking, of course, was felt in northern California, but as a result of the 1971 San Fernando earthquake, the California Department of Transportation enacted new standards for a range of structures, including bridges. And in 1982, retrofit projects were completed on the Golden Gate Bridge and the Bay Bridge to strengthen both bridges should a major earthquake strike nearby. A test of the retrofit came seven years later on October 17, 1989.

Known as the Loma Prieta earthquake because the event originated beneath that peak in the Santa Cruz Mountains, it was the largest seismic event to occur in northern California since 1906. Most people, however, remember it as the World Series earthquake because it occurred just before the start of the third game of the 1989 World Series between the Oakland Athletics and the San Francisco Giants.

Severe damage occurred in the city of Santa Cruz and the surrounding area. There was also a significant amount of damage in San Francisco, especially to buildings with ground-floor garages—so-called "soft" stories—that require open space for cars to turn around and park in. In Oakland, a mile-and-a-half section of a double-decked highway collapsed that was built on mud flats, a soil condition that amplifies seismic motion. Forty-two motorists, most of whom were on the lower level, were crushed when the upper deck fell on the lower one.

The shaking also did damage to the Bay Bridge. The 50-foot spans of roadway on both the upper and lower decks of the Bay Bridge at pier E-9, about halfway between Yerba Buena Island and Oakland, collapsed. At the west end of each span was an expansion joint that the earthquake shaking pulled off a six-inch-wide seat, and at the east end a bolted connection. When the spans were pulled away, they both swung down under gravity, pivoting on the bolted connections, the upper span coming to rest on the lower one. The lower one came to rest on an electrical housing; otherwise, both would probably have swung completely open.

A motorist who was on the bridge at the time of the earthquake saw cars "sliding around like they were on ice." Bruce Stephan, who was driving with a passenger on the upper deck when the span collapsed, would remember thinking, "I'm going through the bridge and into the bay! . . . I knew it was death. I didn't think there was anything that could stop us."

But there was.

His car stopped inches shy of plunging into the bay, the car wedged in the gap between the collapsed upper deck and split-open lower one. He was looking at water. He clambered out of the wrecked vehicle and pulled his passenger with him. They walked a mile to Yerba Buena Island for medical treatment.

The Golden Gate Bridge had no damage. But engineering studies conducted after the earthquake concluded that major sections of both bridges could have collapsed if the intensity of the shaking had been stronger and the duration longer—that is, if a more powerful earthquake—like the one in 1906—had occurred. In particular, such an event might lift the 320-foot arch over Fort Point on Golden Gate Bridge off its bearing pins, causing the entire arch to collapse. That, according to the study, could lead to a chain reaction and possible "collapse of the main suspended span." And it could happen in 60 seconds.

The entire two-mile-long east section of the Bay Bridge from Yerba Buena Island to the Oakland landing has now been replaced with a new bridge. It was more cost-effective to retrofit the Golden Gate Bridge than replace it—and the retrofitting, which would eventually cost a billion dollars, was far enough along by April 2006, the 100th anniversary of the great earthquake, that the bridge no longer faces the potential for collapse. Both bridges have also been covered with a network of sensors that will record how the bridges respond to the shaking of future earthquakes.

The Golden Gate Bridge got a minor test of its new seismic safety on June 28, 2010, when a small earthquake occurred along the San Andreas Fault at exactly the spot where the 1906 quake originated. About 2,000 people felt the 2010 event. According to a bridge engineer, the peak acceleration recorded by any of the sensors on the bridge was 0.5 g, that is, equivalent to half the acceleration of gravity. The engineer characterized it as a "fling." In the 1930s, Chief Engineer Joseph Strauss, using the best information available about ground acceleration during an earthquake—which came from Japan after the disastrous earthquake that devastated Tokyo in 1923—originally designed the Golden Gate Bridge to withstand a peak acceleration of 0.1 g.

The professional disagreements between Lawson and Willis continued after the Golden Gate Bridge was completed, and extended into other topics. It became almost a tradition at scientific meetings that they would disagree and have a shouting match. Whatever position one man took, the other man seemed to take the opposite position automatically.

Fortunately, Willis was wrong about his assessment of the stability of the south pier of the Golden Gate Bridge. He did, however, make an important contribution by reminding Californians that the risk to life and property from earthquakes was, and still is, a continual one. After the 1925 earthquake in Santa Barbara, during which 13 people died and most of downtown was destroyed leaving only a few buildings along State Street still standing, Willis publicly challenged a number of booster organizations—the All-Year Club of Southern California, the California Development Association, the chambers of commerce of San Diego and San Francisco—that tried to downplay the magnitude of the disaster. His effort did lead to the first building codes in California that required earthquake-resistant construction, adopted initially by the cities of Palo Alto and Santa Barbara. It would take another earthquake, in 1933, before such codes were adopted statewide, and then they were required only of school buildings.

Soon after Willis died in 1949, someone visiting Lawson noticed that the Berkeley professor had placed a photograph of his nemesis over his desk. The visitor asked if the photograph was really of Bailey Willis. Lawson just turned to his visitor and smiled.

On March 11, 1949, Lawson made headlines a final time when it was reported that the 87-year-old now-retired Berkeley professor had fathered a child with his 39-year-old wife. Local newspaper reporters found him at the Oakland hospital where the birth took place and asked how this was possible.

"It's nothing," he responded as he passed out cigars. "It happens all the time."

Then, as a diversion, he offered to wrestle any of the reporters. There were no takers.

CHAPTER 5

BLUE CUT AND THE MORMON ROCKS

Our ranch in Valyermo is but a few yards from the Fault
and that really was one reason why we settled there.
—Levi Noble, on living along the San Andreas Fault

To be successful, a field geologist must possess two personal traits: an ability to use any of the five senses to divine the nature of a rock, and a crazed obsession for barren landscapes. If one wonders how the sense of taste might be employed, let me explain.

Years ago, during my one and only class in field geology, I was standing at the bottom of a small basin in the central part of the Mojave Desert near a place appropriately called Calico Peaks. All around, exposed in the walls of the basin, were layers of colors. There were blue mudstones, green siltstones, and buff-colored sandstones. Each layer represented a slow accumulation of sediments at the bottom of a small lake that had disappeared millions of years ago.

On one occasion, the professor who taught the course called the students together. He walked us over to the base of a protruding outcrop near the edge of the basin. He asked us to take up small samples from a thin greenish mudstone exposed on the outcrop and to place the samples in our mouths. We did as instructed.

"Feel the creamy texture," I can still hear him saying, "as you press it with your tongue against the roof of your mouth."

Next, he had us take samples from a dark brown siltstone immediately beneath the green mudstone. This time, he said, we should feel a gritty component as we gently ground the samples between our teeth. Again, we followed his instructions.

When we had finally completed both tasks, the professor, who, as I distinctly recall, had a gleeful tone in his voice, told us that the particles of grit we were now grinding between our molars and swirling in our mouths were fossilized fecal pellets that had been excreted by tiny animals that had once scurried around on the bottom of the lake.

Such is one of the rites of passage to which an individual might be subjected if one wished to be a field geologist. And there are others. Each one, at its core, actually represents a means to identify and characterize rocks in the field.

As to the other necessary trait of a successful field geologist, the desire for barren landscapes is something that I doubt can be acquired but must be inborn. By illustration, I point to the remarkably productive and always self-indulgent career of a New York sophisticate, Levi Fatzinger Noble.

Noble was born into a world of wealth and privilege. His childhood home was a four-storied mansion built in the Queen Anne style complete with steep gabled roofs, large bay windows, and a stone archway over the main door. It was located along the main street of Auburn, New York. Inside were 17 rooms, each with a fireplace, and one of the fireplace mantels, so it was said, had been carved by the hand of Brigham Young when he was 19.

On entering the house, a visitor was struck by the immensity of the rooms. Even the main hall had a fireplace. There was a grand, highly polished mahogany staircase that connected the ground floor to the upper floors. There was a large dining room with a table and chairs that could seat 24, though as far as Noble could remember he had never seen more than six people at the table. It was within such a world of excess that Noble was born in 1882. In many ways, it was a world he never left.

At age 20, still supported financially by his father, a railroad attorney and owner and president of Auburn's largest trust company, Noble entered Yale College. He never gave a reason why he decided to study paleontology—in fact, this shy and unassuming man seldom offered a reason for anything he did—but his interest in fossils and shells led him to geology. And it was that field that would become his passion.

During Noble's sixth year at Yale, still undecided on what to do with his life, he joined a group of students invited by geology professor Herbert Gregory invited students to accompany him on an extended Christmas holiday to Arizona. Each student had to pay his own way, which meant few could afford to go. But Noble was an exception—he always had money. And, though not yet experienced in travel and not yet having lived outside the niceties of a privileged life, he knew the trip would be an easy one because Gregory had just married and this wintertime trip to Arizona would double as his honeymoon. So Noble decided he would go.

The first major stop Gregory, his bride, Noble, and the other Yale students made was at the Grand Canyon. The canyon was not yet the iconic feature it is today. Nor was it yet the mecca for crowds of tourists who now come during all seasons. In January 1908, when the Yale group made its trip, they arrived, as visitors did at the time, at a train station near the south rim of the canyon known as Bass Station. Here, they boarded a stagecoach and rode

20 miles to a conclave of rustic buildings and several worn tents known as Bass Camp, the only accommodations then at the Grand Canyon. Here they could descend along the only known trail that led down to the canyon bottom—the Bass Trail.

In 1908, the entire canyon-rim facility—and access to the canyon—was controlled by one person, William Wallace Bass, a former train dispatcher in New York who had come west in the 1880s for his health and to seek his fortune. His health improved, but a fortune never materialized. Instead, after discovering a trail that led partway down into the Grand Canyon while on a hunting trip, he decided to open a campground, clear the trail, and cater to the occasional tourist. By 1900, he had extended the trail all the way to the bottom. He also constructed a few wooden buildings that served as a kitchen, a dining room, and a recovery room, the last for those bold souls who had risked venturing the full length of the trail down into the depths of the canyon.

Eight years later, the Yale group was making a descent. Though they spent only a few hours at the bottom, one of the students—Noble—came back transformed.

Unlike the northeastern United States where bedrock is often obscured by glacial drift or vegetation and where, for a significant part of the year, the ground is covered by ice and snow, here everything was revealed. It was "the completeness of exposure" and "the certainty of stratigraphic position," Noble would write, that attracted him, "the diagrammatic simplicity" and the fact that the walls of the Grand Canyon were, as yet, "undescribed in absolute detail." And there was another attraction: The mile-wide canyon with giant tiered walls was "lonely and inaccessible," an ideal place where Noble, perhaps for the first time in his life, could find solitude and work alone.

He returned in August 1908 to begin the first detailed study of the walls of the Grand Canyon. For four months he lived inside the canyon, returning to Bass Camp on the south rim only twice, once in September and again in November, both times for a week.

During those months of work, a man named John Walthenburg, who served as canyon guide, mule tender, and camp organizer, also served as his field assistant. Years later, maintaining a close friendship, Noble would hire Walthenburg to work for him after he moved to California.

With no prior field experience—college courses in geology then consisted of classroom lectures and the occasional nature walk when philosophy, not geology, was usually discussed—Noble sampled, studied, and described the many layers exposed in the canyon walls. He also measured the thickness of each layer as it was exposed along the Bass Trail, using a foot-long wooden ruler to make the measurements. For a layer of limestone and shale located near the base of the trail, which he named the Bass Limestone in honor of William Bass, Noble determined the layer consisted of 74 separate units—which included a white cherty limestone and a lamellar blue shale with concretionary structures, both first described by Noble—that had a total thickness, according to his careful measurements, of 304 feet, 9 inches.

One outcome of his work is familiar to anyone who visits and treks into the Grand Canyon today: Levi Noble named most of the layers exposed in the canyon walls.

Before Noble began his work, only one layer had a formal name: the Redwall Limestone, a 500-foot-thick cliff-forming limestone exposed in the upper section of the canyon wall and named by Grove Karl Gilbert in 1875. Names were given to three other layers in 1910—four years before Noble published his work—by Nelson Horatio Darton, a government geologist who had been sent to Arizona to find sources of water for the Atchison, Topeka, and Santa Fe Railway. Darton worked at the Grand Canyon for only two weeks, and to him are owed the naming of the Kaibab Formation, the Coconino Sandstone, and the Supai Formation, all lying above the Redwall Limestone.

Noble named the Hermit Shale, which he found sandwiched between the Coconino Sandstone and the Supai Formation, as well as the Muav Limestone, the Bright Angel Shale, and the Tapeats Sandstone. Deeper in the canyon, he identified the Cardenas Lavas, the Dox Sandstone—which he named after Virginia Dox, an early educator in the American West who had flaming red hair that matched the color of the sandstone—the Shinumo Quartzite, the Hakatai Shale, and, as already noted, the Bass Limestone.

Noble returned to Yale in December 1908, and the next June presented his work to the Yale faculty. For his effort, he was given the college's highest degree, a doctorate. Then, impressed by the thoroughness and quickness of his work, the federal government hired him as a geologist. And he gladly accepted—with an interesting proposal: He agreed to accept a government job if he was paid no salary. That way he would have the standing of a professional geologist—such as Darton and Gilbert—but for the most part he could choose what topics interested him and what he would study, since he would not be at the behest of grants or other economic interests. The federal government readily agreed. And for the next 43 years he worked as a federal civil servant—without pay.

In 1910, Noble was set to return to Arizona and resume his work at the Grand Canyon and the surrounding area when fate intervened. He had fallen in love. And the woman he fell in love with lived in California. It would be she who would direct his extraordinary talents as a field geologist away from the Grand Canyon to a feature that, in the years since 1906, had been fading from scientific interest—the San Andreas Fault.

Levi Noble and Dorothy Evans met through her younger brother, Deane Mann Evans, who was a student at Yale studying metallurgy. The courtship was a short one. On July 7, 1910, Noble and Evans married in Philadelphia, her hometown, and then they traveled across the country to Arizona.

The next two weeks are a mystery—as honeymoons often are; in this case, the newlyweds simply disappeared into the desert. Where they went, what they did, or what they saw was never told. But this much is certain: Dorothy, who had been raised in Pennsylvania and who was now living with her parents in Los Angeles, came away enchanted by the desert. The beauty of a desolate landscape struck her immediately—in the same manner her husband had been struck when he first saw the Grand Canyon—and she came away with a desire to live in a desert environment. So it came as a pleasant surprise, once the newlywed couple reached Los Angeles, that her father offered them the chance to do exactly that.

Dorothy's father, Cadwallader "Doc" Evans, who had made his money designing and patenting safety devices for the steel and iron industries of Pittsburgh, had brought his family to southern California during a series of short stays in the 1890s to escape the northeast winters. By 1900, he had decided they would stay permanently and he was soon heavily invested in real estate. He owned a large parcel close to Long Beach next to Signal Hill, where in 1921 a large oil field would be discovered. Another parcel was located west of Los Angeles among vineyards and barley fields. Once known as Cahuenga Valley, it is today known as Hollywood. A third was a large fruit ranch nestled against the north side of the San Gabriel Mountains and on the edge of the Mojave Desert. Evans asked: Would the newlyweds be willing to accept one of these parcels as a wedding present?

Yes, they would. They chose the fruit ranch because it was remote and on the edge of a desert and—as Levi knew, having read Lawson's report of the 1906 earthquake published just two years earlier—the San Andreas Fault ran through it.

Known as Valyermo, a contraction of the Spanish words *valle yermo* or "barren valley," the ranch was truly barren except for a strip of land right at the base of the San Gabriel Mountains. Here was an impermeable barrier that forced groundwater to rise near the surface, in places within 20 feet, thus providing a plentiful supply of fresh water for the acres of fruit trees. And, as Noble would eventually show, the barrier was a consequence of the San Andreas Fault.

Though both Levi and Dorothy had been raised in extravagant households, they accepted the modest six-room frame house that already existed at Valyermo, adding, at considerable expense, both electricity and a telephone.

Levi also took one of the small storage sheds and converted it into a private study, choosing this particular shed because, as best as he could determine, it lay directly across the trace of the San Andreas Fault. If he was lucky, he would tell friends, he would be at work in his study when the next big earthquake struck and he would be able to see the approach of the ground crack merely by looking out a window.

Most years, the Nobles wintered at Valyermo and summered at the Grand Canyon, where Levi continued to investigate the walls. They also made at least annual trips to the East Coast to live for a few weeks at the mansion in Auburn—which Levi had now inherited and which was under the care of a permanent staff—and to stay in hotels in New York City. While in New York, they rummaged through bookstores for rare books. They had their clothes tailor-made at Brooks Brothers, Levi favoring bright shirts. One design in particular, of which he had several made, was a yellow shirt with red polka dots. He had a suit made up in the manner of Sherlock Holmes, his favorite fictional character, complete with a long cape and a double-billed cap, claiming that this attire was the most practical for geologic fieldwork because it offered both freedom of movement and protection from the sun.

They also favored luxury cars, at one point owning two Jaguars, one kept at the Auburn mansion and the other at Valyermo. For a field vehicle for use in the California desert, the government purchased a Model A modified with extra-low gears and a high clearance. Levi had it painted bright red so that he could easily spot where he parked it in the desert.

During the First World War, Levi was asked by the federal government to survey around Death Valley for nitrate deposits that could supply the munitions industry. As a result, he and Dorothy established a third house in the small town of Shoshone, this one extremely primitive, being constructed of abandoned railroad ties.

His work on the San Andreas Fault began in earnest after the war in 1920 and continued for three years. He focused on a 50-mile segment centered on Valyermo that ran from Soledad Pass south of Lancaster to Cajon Pass north of San Bernardino. It would be along this segment that he would identify many of the more subtle features that would begin to explain what the San Andreas Fault was and how it worked.

As noted in an earlier chapter, before the 1906 earthquake Harold Fairbanks, one of Lawson's former students, had traveled along the same segment of the San Andreas Fault that Noble was to study intensely. Fairbanks made a second trip just months after the earthquake, and with the varied scenes of that event fresh in his mind—having seen and followed the "mole track" and noting, as others did, how it ran along an alignment of hollows and ridges—he saw the same type of features aligned along the remarkably straight boundary between the San Gabriel Mountains and the Mojave Desert.

Noble, of course, recognized the same alignment of features—and much more.

By following the edges of individual scarps, ridges, and depressions, he could identify what he called "the master fault." And bordering the master fault was a zone, in places as much as six miles wide, of parallel branching and interlacing fractures. The zone, in essence, consisted of slivers of blocks or wedges whose long axes paralleled the master fault so that, in his words, "the dominant structure is a sort of slicing."

And within this zone of fractures and wedges he found a curious feature: a belt, a few hundred feet wide, of thoroughly crushed and shattered rock. Close inspection revealed that the rock was actually a powdery granite, evidently pulverized by repeated movements of the fault to form a "fault gouge," the term first used 40 years earlier by miners who uncovered such friable material at the offset ends of mineral veins.

Fault gouge is the bane of road builders. The material has the structural integrity of clay and when mixed with water turns into a soft paste. It clings to everything, is nearly impossible to walk on, and causes heavy equipment to get mired in a muck that seems to have no bottom.

Noble recognized fault gouge along several sections of the San Andreas Fault that he studied. The most spectacular section is a few miles west of Valyermo, where today a highway swings through a series of sharp curves, both to climb a steep slope and to avoid, as much as possible, the soft gouge. Here a badlands landscape has formed—best seen near Appletree Campground—of steep ravines and crumbling knife-edge ridges. The gouge has the feel of talcum powder. It is impervious to water—which is why the San Andreas and other faults act as underground dams, forcing groundwater to the surface and resulting in springs like the one that fed the Nobles' fruit farm. And if one stands back and looks at a large exposure, it is possible to see subtle bands of color that were in the original granite, indicating that this rock is *in situ*—pulverized in place as if it were caught in a vise.

In addition to the zone of fractures and wedges and the belt of fault gouge, Noble recognized yet another key feature of the San Andreas Fault. Along the 50 miles of fault trace that he studied and mapped, he noted that "scarcely anywhere in the fault-zone are the rocks on opposite sides of the master fault similar." The most striking contrast—and one that is easy for a nongeologist to see—is in Cajon Pass at a place known as Blue Cut.

As the lowest gap in a very long trend of east-west mountains, Cajon Pass is the obvious place to run roadways and railway lines.

Because the roadways and railway lines required a low grade, workers had to cut deep into the hillside, revealing in one stretch an astonishing blue-gray, often iridescent rock known as the Pelona schist.

For the fast-paced traveler, the Pelona schist is best revealed in a large roadcut where I-15 runs through Cajon Pass midway between the Kenwood and Cleghorn exits. North of the roadcut the rocks are a tan sandstone. Between the schist and the sandstone runs the San Andreas Fault.

To get a better look at the Pelona schist, one needs to travel the stretch of old Route 66—once regarded as the nation's Mother Road—that runs through Cajon Pass. Still identified by large white stenciled shields painted on the roadway, Route 66 passes through a section known as Blue Cut where restaurants, small stores, and a few motels were once located to cater to travelers during the heyday of cross-country motoring during the 1950s. Route 66 is now all but abandoned, but here at Blue Cut one can find clear evidence of the San Andreas Fault.

There is a large vertical cut of blue rock, the rock cut through with milky-white veins of quartz. Drive north a few hundred feet and the rocks are entirely different: a tan sandstone. Where the transition occurs is part of the San Andreas Fault.

To be even more specific, where the fault trace runs—and to have the opportunity to stand precisely on the trace—one needs to negotiate a dirt road that runs west from Route 66. After crossing three railroad tracks, the road leads to a large parking lot. From there, a trail takes an intrepid traveler to the edge of a small lake—Lost Lake—a feature Noble knew well and described in his reports.

This elongated lake, about 600 feet long and as much as 100 feet wide, is similar to the dozen or so ponds that Lawson and Palache found just south of Mussel Rock—and which are now either dry or filled in—and that led Lawson to conclude that the San Andreas Fault must still be active. If one walks around the south side of Lost Lake, one finds boulders and pebbles of blue Pelona schist. On the north side of the lake is the tan sandstone. The southern edge of the lake marks the fault trace.

To get even more precise, walk to the southeast end of the lake. Here, on the surface, is a perennial mud hole stretched across the landscape and located in an arid environment. What is the source of the water? It is groundwater forced up by fault gouge. That line of mud is the San Andreas Fault.

One can imagine Levi Noble standing here nearly a century ago. His red high-clearance low-geared Model A is parked on the dirt road. He is dressed as Sherlock Holmes, possibly, wearing his favorite yellow shirt with red polka dots. He looks around . . . and sees other indications of the fault.

Far to the southeast is a notch oddly located midway on a hillside. To the northwest is Lone Pine Canyon, long and broad, similar in form to the San Andreas Valley in that Lone Pine Canyon also cuts diagonally across a mountain range. And on opposite sides of the canyon are dissimilar rocks. On the south side is the blue Pelona schist. On the north side is a variety of rock types—granites, gneisses, and even an outcrop of limestone.

All this indicates, with great exactness, exactly where the San Andreas Fault lies amidst this diverse array of geological features. But how does it move? What is the culmination of so many earthquakes?

This is a crucial question. And Noble had an answer.

Early during his study of the fault, Noble was led by a Mr. Peters, whose ranch was at the base of the San Bernardino Mountains just southeast of Cajon Pass, to see a peculiar set of deep ravines. Each ravine had a right-offset dogleg. In each case the amount of offset was about 150 feet.

Taking the 1906 earthquake as typical, when the ground surface shifted to the right about 15 feet, the doglegged offsets at the four ravines seemed to be evidence for ten earthquakes of similar size rupturing along the same fracture.

But, as Noble had determined, the San Andreas Fault had existed and had run along the north side of the San Gabriel Mountains and through Cajon Pass at least since the Quaternary Period—that is, for millions of years. And so, somewhere, there must be evidence for *many miles* of horizontal movement.

And Noble knew where it was.

If one returns to Blue Cut and drives north a mile or so along Route 66 and across the San Andreas Fault, one will see to the west a curious outcrop of high-standing rocks. These are the Mormon Rocks, a series of hogback ridges with deep pockmarks, their unusual shape making them a favorite of movie directors who want to film Hollywood Westerns, and so named because it was here that Mormon pioneers would camp and collect water before crossing the Mojave Desert to Salt Lake City, and where they would arrive after a successful crossing of the desert and have a first chance to replenish their water supplies.

The Mormon Rocks are a thick sequence of tightly cemented sandstone beds, a characteristic that makes them much more resistant to erosion than the surrounding gravel and silt deposits and thus allowing them to stand out in relief. As Noble knew, there was only one other exposure of similarly appearing rocks in the region—and it was within sight of his ranch, 25 miles to the northeast, at Valyermo.

Standing at his ranch, Noble could look south through a notch in the hillsides made by Sandrock Creek at another thick sequence of sandstone beds, also tightly cemented and also with deep pockmarks and high-standing hogback ridges. These are the rocks of the Devil's Punchbowl. And they lie immediately *south* of the San Andreas Fault, while the Mormon Rocks lie immediately to the *north*.

The offset was in the same direction that the ground surface had shifted in 1906 and the same as the four ravines displaced at the Peters ranch north of San Bernardino: a shift to the right. Could it be that the rocks of the Devil's Punchbowl and those of the Mormon Rocks had once been a single continuous geologic unit that had been split and slid apart by the San Andreas Fault so that they were now separated by 25 miles? The similarities were too much to be a coincidence. Noble was convinced. But was the geologic evidence strong enough to convince others?

Though he lived away from and worked outside of established scientific institutions, Noble did correspond with other geologists, often trying to convince them to visit him at Valyermo and see the evidence. On at least one occasion, he wrote to Lawson and offered to pay his travel expenses and said the Berkeley professor could stay at the ranch house at Valyermo. Lawson accepted and brought a colleague—Bailey Willis.

Little is known about their visit except that, one evening after a day of fieldwork and before dinner, Noble asked his guests if they might like a drink and what they would have. As Noble later told the story, Willis replied with a cool detachment. "Nothing at all," the Stanford professor said. "Scotch and soda," was the response from Lawson, maintaining his well-known privilege of always differing with Willis.

But as to the San Andreas Fault and whether it could have accumulated miles of horizontal offset, the professors were in complete agreement: Over the long term, the Earth's crust shifted vertically, not horizontally, raising mountains and lowering basins. The 20 feet or more of horizontal movement during the 1906 earthquake had been an aberration. Future earthquakes along the San Andreas would counter that movement.

As if to reinforce their rare agreement, many years later at a scientific meeting that both attended, the speaker had given what he thought was compelling evidence that there was at least the *possibility* of the Earth's crust sliding horizontally after repeated earthquakes. As one person who attended the meeting recalled, Lawson, who was sitting in the front row, rose to his feet and announced, "I may be gullible! But I am not gullible enough to swallow this poppycock!" Then Willis, who was sitting in the second row, got up and turned to the audience and said, "All you here today bear witness—for this is the first time in 20 years I find myself in complete agreement with Andy Lawson."

And so it went. Noble's evidence was questionable and was dismissed, unsupported by any other observations. Besides, no one had another mechanism in mind to explain how the Earth's crust might slide horizontally for many miles. It had to be the result of earthquakes along the fault.

In 1926 Noble gave a presentation of his work along the San Andreas Fault at a scientific meeting in Japan. It did not go well. By his own admission, it was "a traumatic experience." This fretfully shy man had to stand in front of an audience—this is the only time on record that he ever did so—and describe his work. He had taught himself a few words of Japanese and tried to incorporate them into his presentation with dismal results. There is no evidence that anyone was swayed by his work. And that was just as well, because Noble's primary goal of going to Japan was not to convince others of his conclusion but to feel an earthquake.

Japan is a country of frequent earthquakes. One is felt almost every day somewhere in the island nation. But during the four weeks Noble was in Japan, no earthquakes were felt at the places he visited.

The only seismic shaking of note he did feel during his whole life was in 1933. The event was centered near Long Beach south of Los Angeles, doing a considerable amount of damage and causing more than 100 fatalities.

At Long Beach, the walls of many buildings collapsed, houses were displaced from their foundations, and oil derricks caught on fire. In Pasadena, 30 miles north, people reported a moderate amount of shaking that caused some articles to fall from shelves and cracked the walls of some buildings. At Valyermo, 80 miles north of Long Beach, the shaking was perceptible, though slight.

The shaking occurred during the early evening. Dorothy and Levi were in the kitchen. They quickly turned off the burners and went outside. Dorothy wondered whether this might be the big earthquake that Levi was waiting for, but he said it was not. They would have to wait longer.

Years later, near the end of his life, Levi Noble would tell people that for some reason God had decided to deny him the experience of truly strong seismic shaking. Even the San Andreas Fault, where he had made his home and lived for 55 years, did not produce a significant event.

After Japan, his interest shifted and he spent most of his professional time working along the northern extreme of the Mojave Desert along the Tehachapi Mountains and at Death Valley. He and Dorothy officially made their cabin built of railroad ties at Shoshone their official voting residence, being dedicated Republicans, so that they would offset the strong Democratic majority of the other residents.

In 1965, at age 83, after two years of illness, Levi left California for the last time. He and Dorothy moved into the four-storied mansion in Auburn that had been his childhood home. He gave strict instructions to his staff of servants that he wanted the mansion maintained exactly as it had been when he was a boy, including the placement of his small childhood gloves on a stand next to his bed.

Two months after moving into the mansion, Levi fell and broke his ankle. He was now confined to a wheelchair. One night, Dorothy thought she heard him stirring. She asked if everything was all right. There was no answer. She found him lying lifeless. He had died during the night.

Levi Noble was buried on a hillside that looked over the Auburn mansion. His death on August 4, 1965, came 11 days after the publication of the first scientific paper that proposed the Earth's surface consisted of several large mobile plates. In that paper, authored by Canadian geophysicist J. Tuzo Wilson and published in the British journal *Nature,* Wilson specifically identifies the San Andreas Fault as a plate boundary where there are "large horizontal movements," seemingly supporting Noble's previously mocked claim of 25 miles of horizontal displacement along the fault.

But here is the irony. Noble's conclusion about the San Andreas Fault was *right,* but for the *wrong* reason. The tightly cemented, deeply pockmarked sandstones that stand high as steep ridges—known in geology by the picturesque term "hogback"—as the Mormon Rocks and the Devil's Punchbowl are *not* the same geologic unit and so could *not* have been split and slid apart by the San Andreas Fault. The two sandstones did form in similar environments—deposited by streams that flowed westward from the Mojave Desert—but they did not form at the same time. This came to light in 1972, seven years after Noble's death, when vertebrate bones of the mid-Miocene horse *Merychippus tehachapiensis* were found in the Mormon Rocks, indicating those sandstone beds were at least 5,000,000 years *older* than those exposed at the Devil's Punchbowl, where recovered fossils indicate those rocks were deposited during the late-Miocene about 12 million years ago.

Though Noble was wrong, he did have an influence on the next generation of geologists who would examine the San Andreas Fault in greater detail. They would uncover many geologic units that have been split and slid apart by the fault, determining that the maximum amount of horizontal movement has been *hundreds* of miles, far greater than the 25 miles Noble had originally hypothesized.

But before this next generation of geologists began their work, Levi Noble had a contemporary who was also deeply interested in the San Andreas Fault and in California earthquakes, though in a decidedly different way. His name was Charles Richter.

CHAPTER 6

THE TROUBLED WORLD OF CHARLES RICHTER

I am a dubious work of art.
—Charles Richter, 1976

He was born Charles Kinsinger and did not claim the name Richter—his mother's maiden name—until he was 26 years old, a few years after fate took him into seismology. According to his mother, she and his father married each other twice, each marriage lasting long enough to produce one child and both marriages ending in divorce.

The first one produced a daughter, Margaret Rose, in 1892. The second resulted in Charles Francis, born in 1900. Both brother and sister were beset with mental and emotional problems throughout their lives that eventually led Margaret to live much of her life inside a sanatorium. Her brother seemingly compensated for his own issues by maintaining a lifelong fascination with numbers and calculations and by engrossing himself in the fantastic worlds described in science fiction magazines such as

Other Worlds, Thrilling Wonder Stories, and the ever-popular *Amazing Stories.* Both Margaret and Charles wrote poetry. He became a nudist. And in the public's mind, he is the world's most famous seismologist.

Other researchers contributed more to a fundamental understanding of earthquakes than he did, but it was Richter's accomplishment that has captured the public's imagination. Given the chaos that earthquakes cause and the lingering and unsettled feelings that follow such events, he found a way to quantify and compare—that is, to assign single numbers—to what seems to be a hopelessly complicated yet primordially frightening events. "What was the magnitude?" is the first question invariably asked today after one has heard that an earthquake has struck distant Sumatra or nearby San Francisco or anywhere else. And he developed his magnitude scale, initially, for southern California—for earthquakes along and near the San Andreas Fault.

Born on a farm in Ohio, Richter moved with his mother and sister to southern California in 1909, the family settling in the Wilshire area of Los Angeles. Introverted, as well as shocked by the fast pace of city life and by the combativeness of some children, he found refuge, as many children do who feel lonely and out of place, by going to libraries and reading books. His inclination to mathematics caused him to gravitate toward books about science, an interest that led him to enroll in Stanford University, where he received a degree in physics in 1920. The next year he began graduate studies at Stanford, but left suddenly and returned to Los Angeles, where soon after he had his first major emotional breakdown.

Exactly what the affliction was is not known, but in an engaging biography, *Richter's Scale,* the author Susan Elizabeth Hough, an accomplished seismologist in her own right, explored the possibility that Richter suffered from Asperger's syndrome, an illness that was not recognized and described until 1944, when Richter was well into middle age. Symptoms include discomfort in social situations,

TOP: Andrew Cowper Lawson, who discovered the San Andreas Fault. *Photo courtesy of The Bancroft Library, University of California, Berkeley.* BOTTOM: Passenger train derailed at Point Reyes Station, California, by the 1906 San Francisco earthquake. *Photo courtesy of the United States Geological Survey/Gilbert.*

TOP: View down Sacramento Street, San Francisco, April 18, 1906. Notice the collapsed walls on the buildings at the right. *Photo courtesy of The Bancroft Library, University of California, Berkeley.* BOTTOM: View down Sacramento Street, San Francisco, 2006. *Photo by the author.*

Alice Eastwood standing along the "mole track" formed by the 1906 earthquake. *Photo courtesy of the United States Geological Survey/Gilbert.*

TOP: The Lawson house in the Berkeley Hills. *Photo by the author.* BOTTOM: The Hayward Fault passes directly beneath the football stadium at the University of California at Berkeley. *Photo courtesy Google/Seismological Laboratory, Berkeley.*

TOP: Charles Richter in 1927. BOTTOM: Charles Richter in 1979. *Both photos courtesy of the Archives, California Institute of Technology.*

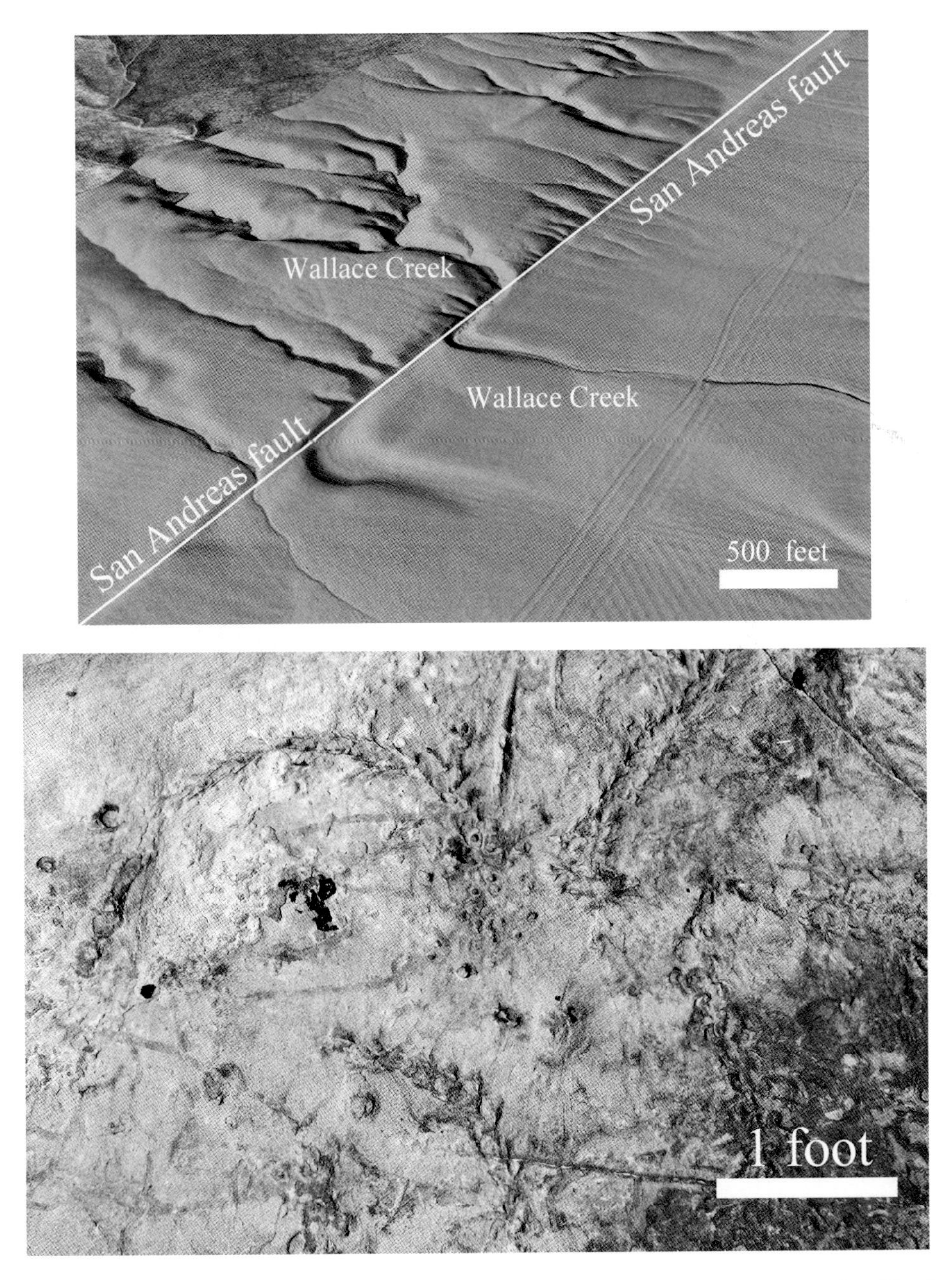

OPPOSITE TOP: Collapse of the roadway on the Bay Bridge that connects San Francisco and Oakland, Loma Prieta earthquake, October 17, 1989. *Photo courtesy of Charles E. Meyer, copyright 1989.* OPPOSITE BOTTOM: Destruction in the Marina District of San Francisco, Loma Prieta earthquake, October 17, 1989. *Photo courtesy of John K. Nakata, copyright © 1989.* TOP: Aerial view of stream offsets along the San Andreas Fault, Carrizo Plain, California. Notice the right-step offset of Wallace Creek near the center of the photograph. *Photo courtesy of OpenTopography/Crosby.* BOTTOM: Fossil imprints of *Hillichnus lobosensis* at Weston Beach, Point Lobos, California. *Photo by the author.*

TOP: Offset sidewalk caused by creep along the Calaveras Fault, Hollister, California. *Photo by the author.* BOTTOM: Lion Gate at the ancient city of Mycenae, Greece. The solid rockface to the left is a scarp produced by a prehistoric earthquake. *Photo courtesy of Andrew Selkirk.*

physical clumsiness, and an obsession with a narrow subject. Such characteristics are not uncommon among those who excel in scientific research, but for those who have Asperger's syndrome, the behavior is extreme, considered by some to be a mild form of autism. Richter suffered the additional symptom of episodes of uncontrolled and unexplained weeping.

Back in Los Angeles, Richter put himself under the care of the same psychiatrist who had treated his sister and he voluntarily lived for a time at a sanitarium. After a year, he left and found a job as a messenger for a museum in Los Angeles. Later, he went to work as a clerk in a hardware warehouse. It was during that time, in 1923, that he heard that Robert Millikan, the new president of the California Institute of Technology in Pasadena and a recent recipient of a Nobel Prize in physics, was giving a series of public lectures in which he would describe the experiments that had led to the award. Richter decided to attend those lectures.

The lectures reenergized Richter's interest in physics and mathematics, and he soon submitted an application to the institute and was admitted to the graduate program. In fact, it was Millikan who suggested a subject for him to study—to use the new ideas of quantum mechanics, which had its roots in a scientific paper contributed by Albert Einstein in 1905, to calculate the behavior of a hydrogen atom with a single spinning electron. As Richter was nearing the end of his work, for which he would receive a doctorate degree, Millikan called him into his office.

The institute's president asked the soon-to-be graduate whether he might be interested in staying at the institute and working in a different field. A seismological laboratory had recently been organized and earthquake records were accumulating at a fast pace. Would Richter be willing to examine and catalogue those records?

The 23-year-old did not hesitate: Yes, indeed, he was interested. And so a crucial element in the history of seismology was born.

The first *sismografo,* or seismograph, an instrument that could record on paper the seemingly erratic motions of the ground during an earthquake, was built in the 1850s by Luigi Palmieri, then the director of Vesuvius Observatory in Italy. His design was something of a Rube Goldberg affair—it consisted of a series of complex gadgets including two U-shaped glass tubes containing mercury, a pair of electromagnets, a few coiled springs, a host of levers, a pen and a roll of paper, and a pair of pendulum clocks that worked in sequence—an instrument complicated by today's standards, but one that did produce some of the first permanent records of earthquake shaking. To set up Palmieri's *sismografo* was also complicated. Each element had to be carefully adjusted to move whenever the ground started to vibrate, even feebly.

To prepare it to record an earthquake, one clock was set to the correct time and its pendulum set to swing, while the other clock and its pendulum were stationary. Then when the ground shook, the level of mercury in each glass tube changed. That change completed an electric circuit that sent power to the electromagnets that caused sequences of levers to move. One sequence stopped the pendulum of the running clock, thereby establishing the *time* of the earthquake. Another sequence of shifting levers started the pendulum of the second clock to swing, the swinging of the pendulum continuing until the ground shaking stopped. That caused the mercury to return to its original level, which opened the electric circuits, which cut off the power to the electromagnets, which caused the levers to move back, which stopped the swing of the second pendulum, thus allowing someone to read, from the second clock, the *duration* of the earthquake.

The second clock also had another purpose: As soon as the pendulum started to swing, a spring was released that unraveled a roll of paper. And as the paper unraveled, it ran under a pen that etched a permanent trace of the ground shaking.

In all, it was a hopelessly complicated instrument that needed constant adjustment and frequent repairs, though Palmieri claimed that, on April 23, 1872, his *sismografo* "became agitated," and a day later lava gushed from the main crater and reached the foot of Vesuvius in a couple of hours. It was a clear forewarning of an eruption by recording earthquakes, but Palmieri's instrument was never used widely and has no modern descendants.

The direct ancestors of today's seismographs—properly called "seismometers" because the emphasis has been on devising a means to "meter," or measure, ground motion—are a series of instruments designed and built by John Milne, a British mining engineer who in 1876 arrived in Japan to teach at the Imperial College of Engineering in Tokyo. On the night of his arrival in Japan, he felt his first earthquake ever. From that, he became fascinated by seismic shaking and how it might be recorded and what could be learned from it.

Milne's instruments relied on the same basic principle: the movement of a horizontal pendulum. In his most popular and successful instrument, he placed one end of a rigid horizontal bar against a vertical post, leaving the other end free to swing, similar to how a door swings on its hinges. When an earthquake hit and the ground shook, the horizontal rod swung back and forth, causing a pen attached to the free end to trace out a record of the ground motion on a piece of paper attached to a slowly revolving drum.

A Milne-type instrument was the first type of seismometer to operate in the United States, installed at Lick Observatory in California in 1887. The first earthquake it recorded was a "slight shock" on April 24, 1887. The time of the earthquake was given in the logbook at Lick as "night."

By the 1910s, Milne-type seismometers were in use at more than 100 sites around the world. Most were in Japan and Europe. Instruments were also being used in Shanghai, China, in Havana, Cuba, and Manila in the Philippines. Others were operating in Sitka, Alaska, and in the Canal Zone in Panama. Colleges in the United States had them installed in Albany, New York; in Baltimore, Maryland; in Cambridge, Massachusetts; and of course in Berkeley, California. Several were maintained by the Carnegie Institution in Washington, D.C. And one was in operation at the top of a volcano in the Hawaiian Islands. But all of these instruments had major drawbacks: They were expensive and heavy—a single instrument cost about $1,000 and weighed a few hundred pounds—and because of the design they could only record relatively slow oscillations, and therefore could only be used to study seismic waves that arrived from distant earthquakes. In order to record the rapid oscillations of local earthquakes, and thereby be able to locate them accurately, a new design was needed.

One was successfully developed at Millikan's California Institute of Technology and was operating by 1921. Known as the Wood-Anderson seismometer, it was named for its two inventors, Harry Oscar Wood and John Anderson. Wood had worked for Lawson after the 1906 earthquake, producing a map of shaking intensity based on the severity of damage found throughout the Bay Area. In the 1920s, supported by the Carnegie Institution, he founded and was the first director of the Seismological Laboratory, located in Pasadena. Anderson was a designer and builder of astronomical equipment for the new telescopes then being constructed at Mount Wilson north of Pasadena. Together, they produced an instrument that was small, lightweight, and inexpensive to build—and that could record local earthquakes.

Their design was still based on a horizontal pendulum, though of very small size. A thin wire of tungsten was stretched tightly between the ends of a narrow metal cylinder. Attached to the wire was a small ball of copper. When the ground shook, the cylinder vibrated, causing the ball of copper inside the cylinder to twist back and forth on the wire.

It was a remarkable feat of engineering, foremost because even a tiny amount of shaking would cause the copper ball to twist the wire ever so slightly.* A Milne-type instrument could magnify ground motion about 100 times. A Wood-Anderson instrument, because it was lightweight and delicate (the ball of copper weighed only three-hundredths of an ounce, about one gram), could magnify ground motion more than 2,000 times! And the motion was recorded on a roll of paper, similar to the Milne-type instruments.

When Richter began working at the Seismological Laboratory in 1927, four Wood-Anderson seismometers were in operation in southern California—one in Pasadena, another in Riverside, and two along the coast at La Jolla near San Diego and at Santa Barbara. Richter's job was to examine a day's record from each station, noting when the first disturbance of an earthquake had arrived at each station, as well as any other notable characteristics of the seismic waves. He got so good at recognizing wave trains that once, late in life, inspecting a record, he could tell a colleague, "This seems to be exactly the same seismogram as one I saw five years ago."

The arrival times for each earthquake were then used to compute where the earthquake had originated. Richter also wanted to give some indicator of the size of an earthquake, initially planning to list it as "small, intermediate, or large." But he abandoned that idea when in 1933 an earthquake struck south of Los Angeles at Long Beach and there was the mistaken impression by the public, because many buildings had been damaged and some fatalities had occurred, that this had been "a great earthquake," comparable to the 1906 event.

---

* This, of course, produced a new challenge: to isolate such delicate equipment from the normal small vibrations of everyday life, such as a person walking through a building or a vehicle passing nearby. So seismometers are now installed in concrete cellars or down boreholes. The first Wood-Anderson instrument was installed on a concrete floor in the basement of the Mount Wilson Observatory in Pasadena. It was so sensitive that it recorded the rapid pulses of an earthquake that originated in Japan and that devastated Tokyo on September 1, 1923.

Richter, examining the tracings from the seismometers, knew different. But how could he convey this in a simple manner to the public, who had witnessed such visible signs of physical destruction?

The Long Beach earthquake struck on Friday, March 10, 1933, at five minutes to six in the late afternoon. Richter was still in his office, planning to attend a meeting of a local chess club later that night.

He happened to be looking in the direction of one of the revolving drums where an ink pen was tracing out the signal from one of the seismometers and noticed that the pen started to move just as the seismic wave hit the building. As he would recall, he made "appropriate remarks"—one assumes these were directed at the shaking and contained expletives—then moved closer to the drum.

The pen was now swinging wildly. All that told him was that a powerful earthquake had struck somewhere near Los Angeles. Then the telephones started to ring.

By talking to callers, Richter could determine that the earthquake was centered somewhere south of Los Angeles, a conclusion borne out by later radio newscasts. In fact, the greatest damage was at Long Beach, where practically every building in the business district was left in shambles. Here and in nearby communities, more than 100 people were killed.

Notwithstanding the extensive damage and the death toll—the earthquake had originated almost directly beneath Long Beach—from a seismic perspective this had been a moderate earthquake, not a major one. And the idea, quickly voiced in newspaper editorials, that the people of southern California could now relax because no more important shocks need be expected for many years was a false one. As Richter would later write in his textbook *Elementary Seismology*, "A small shock perceptible in the Los Angeles metropolitan center will set the telephone at the Pasadena laboratory ringing steadily for half a day, while a major earthquake under some remote ocean passes unnoticed except for seismograph readings and rates a line or two at the bottom of a newspaper page."

How to change that perception? For the past year or so, Richter had been working on an idea, first proposed by Kiyoo Wadati in Japan, to use the maximum ground displacement during an earthquake as a measure of its size. But there was a lot of variation here: Earthquakes occurred at different distances from seismometers, the ground responded differently depending on the nature of the surface—loose sediments shook significantly more than solid rock—to name just two, and a strong earthquake in an unpopulated region would be reacted to differently in the hearts and minds of the public than a much weaker one that occurred in a large populous city.

Richter at least had a way to account for the physical variations: He had access to the densest network of seismometers and to the largest number of seismic records in the world. And earthquakes were being recorded at a rate of about 100 events per year on the network of Wood-Anderson seismometers. So he had an abundance of data to study.

And, just as important, he reveled in hunching over a table with a pencil, a ruler, and a pad of paper, measuring his seismograms. People often found him working alone in the measuring room, talking to himself. In fact, he seemed capable of carrying on long conversations with himself.

He also loved to calculate. Numbers clearly soothed him. Among his professional papers, kept at the archives at the California Institute of Technology, is a box filled with his search for prime numbers, which is a number that is divisible only by itself and by the number one.

But his mood was unpredictable; it was not possible to judge what might inflame him. On one occasion, recounted in Hough's *Richter's Scale,* he stormed out of a meeting and slammed a door behind him, shattering the glass pane. After a few seconds of dead silence in the room, Richter returned and informed his colleagues, "I am not sorry about what I said but I am sorry about the door."

Nevertheless, during those long, solitary days he found a way to determine the size of earthquakes, and from that he was able to propose his famous earthquake magnitude scale.

There are many misconceptions about Richter's magnitude scale. First, it is not a physical device. There is no specially inscribed ruler or other measuring instrument. As Richter often stressed, it is a *method*, and so it was not invented but devised.

Second, the magnitude scale is arbitrary—that is, it does not have any physical meaning. In this regard, the magnitude scale is similar to the much more familiar Fahrenheit (or Celsius) temperature scale, which is also arbitrary. Both the Fahrenheit temperature scale and Richter's earthquake magnitude scale are based on a specific way to measure temperature and ground displacement, respectively. The Fahrenheit scale, developed in 1724 by Daniel Gabriel Fahrenheit, relied on the use of a mercury thermometer; the earthquake scale, as noted, relied on a Wood-Anderson seismometer.

Both also have arbitrary zero points. For the temperature scale, 0°F was the lowest temperature Daniel Fahrenheit could achieve by preparing a mixture of water and salts. For the earthquake scale, Richter decided that an earthquake of magnitude zero would correspond to the smallest displacement a Wood-Anderson seismometer could measure—0.00004 inch, or one micron, about one-hundredth the diameter of a human hair, if the seismometer was 62 miles, or 100 kilometers, from the earthquake.

As to establishing the interval size of each scale, for the temperature scale, Daniel Fahrenheit, for a reason still debated by science historians, said the boiling point of mercury was 600°F, probably to make the temperature of the human body—98.6°F—close to 100. As to the earthquake scale, Richter realized by inspecting his seismograms that ground displacements during earthquakes could range by more than 1,000,000, and so the earthquake scale became a *logarithmic* scale. That simply means that, for every increase of ground displacement by a factor of ten, the earthquake magnitude increases by one unit. This is easiest to understand with an example.

If the maximum displacement produced by an earthquake at a distance of 62 miles, or 100 kilometers, is one micron, then, as already noted, the Richter magnitude is zero. If the maximum displacement at the same distance is 10 microns, then the magnitude is one; if the maximum displacement is one millimeter—about one-twentieth of an inch—the magnitude is three; and so forth.

According to this scale, if the maximum ground displacement at a distance of 100 kilometers is 100 millimeters—about four inches—then a magnitude-5 earthquake has occurred—and, from experience, many people may have felt the earthquake and there may be some damage.

It is important to note that there is no upper bound to an earthquake magnitude: A magnitude-10 earthquake is possible, but it does not happen because of the material the Earth is made of and because of the planet's size; it is, as Richter often said, "a limitation of the earth, not in the scale."

Richter formally announced his earthquake scale in 1935. It took a few years for the seismological community to adopt it, but it remained essentially unknown to the public until 1952 when the first strong earthquake since his development of the scale struck California. It occurred in an agricultural district south of Bakersfield in Kern County along the White Wolf Fault. This earthquake was strong enough to shorten a railroad tunnel by three feet, leaving the rails bent into S-shaped curves. A study of his seismograms and some calculations showed that it was a magnitude-7.5 event. Because of its location, it did little damage—a few broken windows and cracked plaster, though 12 deaths occurred when a building collapsed. By comparison, the 1933 Long Beach earthquake was a magnitude-6.3 event and, because it occurred in a highly populated area, it devastated several communities and killed more than 100 people.

Since 1952, Richter magnitude has been a staple of earthquake reporting. Moreover, the phrase "off the Richter scale" has come to mean that something is too extreme to measure or so awesome that it is unbelievable. Political commentators use the expression to indicate when an election campaign is in calamity. It also appears in slang, as in "Dude, we were boardin' yesterday and the waves were off the Richter scale!"

The calmest period in Richter's life, when he seemed free of nervousness and other anxieties, was the years he spent devising his magnitude scale. It was at the end of this period—when he introduced his scale—that he and his wife of seven years, Lillian, began to join nudist colonies.

Lillian and Charles had married in 1928; he later mentioned to friends that she was "a little wild" at the time. The nature of her "wildness" he never explained, though at the time she met Richter Lillian was married, had a son, and was separated from her husband. Her attraction to nudism she never explained.

His attraction is clear. One might expect that, because as a youth he had gone on numerous day hikes to the nearby San Gabriel Mountains and on overnight trips to the southern Sierra Nevada Mountains, his attraction to nudism came about as an extension of his outdoor activities. But it was not. It was about friendship.

"I might say," Richter would write, "that I have never known what real friendship was; I have never had any intimate friends—until we joined the Glassey group."

The Glassey group referred to psychotherapist Hobart Glassey and his wife Lura, who ran the Fraternity Elysia, a nudist organization located in a small canyon along La Tuna Canyon Road midway between Pasadena and the San Fernando Valley. It was one of the first nudist camps in the United States, opening in 1934. Here Richter was no longer a professor, nor was he known as Charles;

instead, he was called "Charlie." It is unknown whether he shared with other members of the group anything about his professional life. Whatever it was about this nudist colony, for "Charlie" social barriers and anxieties came crashing down along with clothing, allowing him to connect socially in a way he never could before, even amongst his scientific colleagues. Nor is it known how long he and Lillian remained members—the Fraternity Elysia closed in 1946—though in later years, according to family members, Richter and his wife apparently continued to practice nudism on their own during backcountry hikes.

He was, as many people would say whether the encounter was brief or they'd known him for years, simply *odd*. A case in point was the impression he made on two strangers with whom he and Lillian shared a mule tour of the Grand Canyon. "At dinner we noticed what a peculiar person 'Charles' was," wrote one of the mule riders. "He never looked at you when he spoke and just sat with a grin on his face that seemed to indicate he was in on some joke that you were not." Later during the tour, the same writer speculated on what Richter might do for a living, suggesting facetiously "maybe he just caught butterflies," then concluding that "he was too slow for that." They never learned how brilliant he was!

Among his professional colleagues at the Seismological Laboratory, one, when he was first introduced to Richter, thought, "I had met a hobbit with bushy eyebrows." Photographs spanning several decades do show that he had a lopsided, disheveled look about him, augmented by two tufts of hair that stood out from opposite sides of his head.

Another colleague, who worked closely with Richter for many years and who obviously chose his words carefully, said of the famous seismologist, "He could be charming or irascible; he could be outgoing or shy; he could be gentle and warm or abrupt and cold. And he was a man with a truly remarkable memory, but at the same time was renownedly absentminded." He was also bumbling.

Once, during lunch, Richter tapped an egg against a counter, expecting to peel it. Instead, it broke and the contents ran down the table toward Richter's boss, John Buwalda, the head of the geology department at the California Institute of Technology. Richter had thought he had brought a hard-boiled egg from home that morning.

At times, he exhibited a noticeable facial twitch and always worked with bird-like nervous energy. But he also showed tempers and tantrums. Bursts of rage could occur at any time, leaving those who witnessed such displays puzzled as to their cause. One person who knew him well wrote, "Many of us have fallen under his wrath at times."

He was also a man of quick wit. To recall one such example, he was a guest on a radio talk show and the subject, as expected, was earthquakes. One female caller asked, with intense anticipation, "Oh, Dr. Richter, I'm so afraid of earthquakes. What should I do?" Richter replied, "Get the hell out of California!"

According to Richter, the original intention of the magnitude scale "was to take some of the nonsense" out of earthquake studies. In this regard, he realized his work was incomplete. To be meaningful, earthquake magnitude had to be related to a physical quantity, such as energy. In 1942, he and Beno Gutenberg proposed such a relationship, though they had no way to test it until the era of nuclear explosive testing.

The first nuclear test, which was known as "Trinity" and was exploded near Alamogordo, New Mexico, on July 16, 1945, sent out seismic waves that were recorded on most of the Wood-Anderson seismometers then installed in southern California.* Because the

* Equipment designed to time the Trinity test failed to operate correctly, so the exact detonation time—as now recorded by history—was deduced by Gutenberg and Richter from seismic records to be at 05:29:21 Mountain Standard Time on July 16, 1945, plus or minus two seconds.

device—known as "the gadget"—was detonated in the air, there was no way to know how much of the energy of the nuclear blast had gone into producing seismic waves, so this first nuclear explosion could not provide a test of the Gutenberg-Richter energy-magnitude relationship. Likewise, the second controlled nuclear test, "Able," conducted on Bikini Atoll, part of the Marshall Islands in the Pacific Ocean, was also detonated in the air. The third test, "Baker," also conducted on Bikini, was fired 90 feet underwater. Seismic waves from that explosion were recorded by instruments around the world—in California, the seismic waves were stronger from the Baker test than from Trinity—and gave Gutenberg and Richter their first chance to test their energy-magnitude relationship. It failed miserably.

The seismic waves arrived, as expected, about eleven minutes after the Baker test. From the maximum amplitude of the waves and knowing the distance between California and the test site—about 4,500 miles—Gutenberg and Richter estimated the strength of the Baker test to correspond to a magnitude-5.5 earthquake. According to their proposed magnitude-energy relationship, the amount of seismic energy released by the blast was equivalent to 250 kilotons of TNT. But the actual energy of the explosion—computed from the yield of the nuclear reactions—was much less, only 23 kilotons. That forced them to revise radically how they estimated seismic energy from earthquake magnitude. Another decade of work was required, using the seismic shaking of other nuclear explosions, before Gutenberg and Richter arrived at what is now regarded as an accurate way to convert magnitude to seismic energy.

As a rule of thumb, a one-kiloton explosion, whether by a nuclear device or by a conventional chemical bomb, equates to a magnitude-4.0 earthquake. Furthermore, an increase of one earthquake magnitude corresponds to a 32-fold increase in energy. That means that the largest nuclear device detonated by the United States—the Cannikin underground test on Amchitka Island, Alaska, in 1971, which had an energy yield of 5,000 kilotons—corresponded, in energy release, to a magnitude-7.1 earthquake. Not an inconsequential event, but 1,000 times smaller than the largest earthquake yet recorded, the Chilean earthquake of 1960.

And that shows how truly powerful the largest earthquakes are. As an example, the 2011 earthquake that originated near Mineral, Virginia, which damaged the top of the Washington Monument and caused buildings to sway in New York City and was probably felt by a third of the population in the United States, released about 500 kilotons of energy. The most costly earthquake to date in the United States, the 1994 Northridge earthquake in the San Fernando Valley, which caused $44 billion in damages, was 24 times more powerful than the 2011 Virginia earthquake. And the 1906 San Francisco earthquake was 64 times more powerful than the Northridge quake and 1,500 times the Virginia quake—numbers that should give people charged with responding to earthquake disasters pause that a truly colossal event such as a repeat of 1906 has not struck the contiguous United States in more than 100 years.

Throughout most of his life, earthquakes were at the core of Charles Richter's existence. In a strange way, they also followed him after his death.

The first earthquake shaking he remembered occurred when he was ten and living in Los Angeles. His later work would show that it was a magnitude-6 event that occurred on May 15, 1910, along the Elsinore Fault southeast of Los Angeles. The last shaking he probably ever felt was during the early morning hours of August 4, 1985, also a magnitude-6 event, this one originating north of Los Angeles near Coalinga. He died seven weeks later on September 30 at the age of 85.

Between those two events, he felt scores of earthquakes, though most were slight shakings. But it was the Northridge earthquake on

January 17, 1994, in San Fernando Valley northeast of Los Angeles that limits our ability to ever understand him.

His nephew Bruce Walport lived in Granada Hills near the earthquake's center. He had inherited many of his uncle's prized possessions, including rare books of science, art, and literature. Walport also had family home movies and diaries. He thought he was prepared for an earthquake, having stocked his house with emergency food and insured his house with an earthquake policy.

But minutes after the Northridge earthquake hit, fires broke out where gas mains had broken. One of the houses that burned was owned by Walport. So there is irony to the life of Charles Richter: The personal possessions of the world's most famous seismologist—things that could have given insights into the man—were lost, years after his death, in the aftermath of an earthquake.

CHAPTER 7

OF PETROL AND PINNACLES

I got familiar with rocks, just like a rancher
gets familiar with his cattle.
—Thomas Dibblee, on mapping
the San Andreas Fault

Though it is hard to believe today, during the early history of the state of California, the abundance of oil was a nuisance. It fouled water supplies. Cattle roaming in pastures became mired in tarry pits. In some sections of a young Los Angeles, wagon drivers passing over dirt roads had to stop repeatedly to scrape away a black goo that stuck to the wheels, impeding their progress.

The first person who recorded a visit to California, Gaspar de Portolá, encountered the black goo. In July 1769, after feeling several seismic shakings, he was passing through what would later become known as the Wilshire district when he noticed several large pools of slowly bubbling black pitch. He wondered whether "this substance which flows melted from underneath the earth could occasion so many earthquakes." Two hundred years later, geologists, after studying the situation carefully, would show that indeed there is such a connection, and a surprising one at that. Though long before that connection was made, Californians had changed their minds about the foul-smelling wheel-clinging oily tar that fouled water supplies, and they sought it out in earnest.

The first production of oil in California was in the northern part of the state in Humboldt County, where in 1865 a deep pit was dug at a settlement then known as New Jerusalem. Oozing from the pit was a dark gooey substance that the citizens started to send to San Francisco to cover roofs. To celebrate the discovery—and the economic gain it was expected to bring—the citizens of New Jerusalem rechristened their town Petrolia,* a name derived from "petroleum," meaning "oil rock." But the bonanza was short-lived—the oil extracted at Petrolia was heavy in tar and had little commercial value.

A commercially viable product was finally found decades later in downtown Los Angeles. In 1892, a stranded miner named Edward Doheny noticed, as many had already, that tar often clung to passing wagon wheels. He tracked it back to the source, a pit that today would be at the corner of Colton Street and Glendale Boulevard, about a half mile southwest of Dodger Stadium. He and a prospector friend, Charles Canfield, hurried to lease a nearby city lot and they started to dig. At 50 feet, they struck a pocket of gas and were almost asphyxiated. They hired an experienced driller who, at a depth of 600 feet, brought in a well of crude oil that was soon producing 50 barrels of liquid oil a day. It was a fast-flowing liquid crude—which meant it had almost no solid particles of tar in it and would burn easily—so they sold it as a cheap alternative fuel for kerosene lamps. They made a fortune.

Others followed suit. Within five years, nearly 3,000 oil wells had been drilled in and around Los Angeles. The first California

* Petrolia enters the story of the San Andreas Fault in another way: It is the northernmost community to sit astride the fault and its few buildings were totally demolished by the 1906 earthquake.

oil companies were formed. So much crude oil was being produced that a few enterprising souls realized that a new and much bigger market than as a kerosene alternative was needed if they were going to continue to make money. One of those enterprising souls was Wallace Hardison, a founder of Union Oil.

In 1894, Hardison arranged with executives of Southern Pacific Railroad to view a demonstration of how much more efficient an oil-burning locomotive was than a coal-burning one. Hardison stacked up a line of railroad cars behind an oil-burning locomotive, then had the locomotive pull the cars up through steep Cajon Pass. The executives were impressed. Hardison then showed them the numbers: Oil was cheaper than coal. After that demonstration, the mania for California oil began, and technology dependent on oil was soon to follow.

By the early 20th century, automobiles had become popular—and there was a search for more oil fields. Additional ones were discovered south of Bakersfield and at Huntington Beach and Signal Hill. By 1925, California was producing almost a quarter of the world's oil—and the production soon outstripped demand. Oil executives decided to fix prices, but before they could act, in 1929 the nation's economy collapsed. The demand for oil fell dramatically.

The Second World War and the postwar economic boom fueled a new mania for California oil, but by then all of the obvious places to drill—oil seepages, tar pits, natural gas leaks—had been explored. So a new strategy was adopted.

In the 1920s, if a college-educated geologist showed up at an oil site he was apt to be sprayed intentionally with oily mud by workers and referred to contemptuously as a "mudsmeller." But that was to change. By the 1940s, college men were welcome in the oil fields because they brought a new way to find oil: They would use the principles of geology to decide where the black gold might still be hidden.

But where should they look? If one examines a map of the major oil fields of California up to the 1930s, one notices that the most productive fields occur in one of three large regions: They lie west and south of Los Angeles and in nearby Ventura County, in the Coast Ranges around and north of Santa Barbara, or in the southern half of the San Joaquin Valley south of Bakersfield. At the center of these regions is a vast area where no one had yet found oil—and experienced oilmen could tell you why: It was cut through by the San Andreas Fault. The fault, so it was argued, had mangled and squeezed and disturbed the rocks so that oil could not accumulate in reservoirs.

But a few upstart college-trained geologists thought otherwise. One of them was a shy, lanky, often clumsy individual who for many years could be seen walking alone across the California landscape.

Thomas Wilson Dibblee Jr. was the quintessential Californian. His mother, Anita Orena, was a great-granddaughter of Captain José Antonio De la Guerra y Noriega, who in the early 1800s was appointed by the Mexican government to be the commandante of the Presidio at Santa Barbara. Dibblee's father traced his family's New World ancestry to Ebenezer Dibblee, an Englishman who immigrated to Massachusetts in 1635 and whose descendants continued west and arrived in California with other gold seekers in 1859.

Young Thomas Dibblee spent his childhood on a 20,000-acre ranch, Rancho San Julian, part of an original Mexican land grant located northwest of Santa Barbara at the foot of the Santa Ynez Mountains. At age 18, during his final year of high school, he was sent by his father out across the ranch with a consulting geologist—who, as chance would have it, had been educated by Lawson at Berkeley—to assess the potential for oil. No oil was found, but from the experience Dibblee began a lifelong interest in fossil collecting and gained an immediate appreciation for what could be learned by seeing how rocks are situated and how they are layered.

In 1936, Dibblee graduated from Stanford University—taught, in part, by Bailey Willis—with a doctoral degree in geology and paleontology. He worked for a short time for the state of California inventorying mercury deposits but, bored with a desk job, soon found work preparing geologic maps that could be used to search for oil.

The company that hired him was the Richfield Oil Company of California. Formed in 1905, during its first three decades of existence Richfield went through several bankruptcies and court-ordered reorganizations and mergers with other failed oil companies. In the 1940s, it was still struggling to find a major oil reservoir—most of the reservoirs then in production in California were under the control of a few big oil companies such as Standard Oil or Union—and the executives of Richfield decided they would find one by searching for it in a radically different way. First, the Richfield executives hired a cadre of recent college graduates such as Dibblee. Then they ordered the graduates to look for oil where no one had yet found it—in the vast region centered among the major known fields and cut through by the San Andreas Fault.

In this region of the Coast Ranges—the mass of mountains that parallel the coast of southern Oregon and northern California as far south as Santa Barbara—are a number of elongated basins that seem ripe for oil prospecting. Each basin had been explored by the big oil companies and several exploratory wells had been drilled, but no one had ever been able to find an oil reservoir and establish a producing well. That fact led many to believe that, somehow, the San Andreas Fault had prevented the accumulation of oil. Nevertheless, with few other alternatives, executives of the Richfield Oil Company directed its college-trained men to search in this region. And so Thomas Dibblee went out to see what he could find.

His method of doing fieldwork still makes other geologists cringe. On a typical day, Dibblee slept until midmorning. Once awake, he would walk around for hours and return to camp and, after a quick meal of hard biscuits and coffee, he would sleep again.

A day's main work would begin in the midafternoon, when he would start a long walk. And Thomas Dibblee was a prodigious walker, typically covering 10 miles or more a day. Instead of a backpack, he carried a paper shopping bag, which was not replaced until after he married and his wife sewed a cloth one for him. Inside the bag were the essentials for his work: maps, pencils, a compass, rain gear, a water bottle, and a large box of raisins to support himself on his treks. Instead of wool socks and heavy boots, which most geologists consider indispensable in their work, Dibblee favored thin socks and worn-out sneakers so that he could move quickly and cover as much ground as possible.

He seldom took rock samples, which is another thing that infuriates geologists who try to follow his work today. Those few who ever accompanied him in the field and could keep up with his pace cannot remember ever seeing him chip away at a rocky outcrop to see what a newly exposed surface might look like. And he seldom wrote in a notebook, which many colleagues considered a sin. And he never stood still. He walked and walked, constantly drawing on his map while still moving forward, tracing out, as thin penciled lines barely visible on his map to anyone except himself, the geology that he was walking past. In fact, he covered so much ground and drew maps in such detail that it is said if you drive any road that connects San Francisco and Los Angeles and look at any distant outcrop of rocks, even those on steep peaks, it is almost certain that Thomas Dibblee once stood there.

His nighttime routine was almost as unconventional as his daytime one. He seldom returned to camp before darkness—his camp defined by wherever he had left his car, a Ford coupe. Then he had his major meal of the day: a slice of bread, a half can of beans, a few leaves pulled from a head of lettuce, several cups of water, and finally a cup of coffee. He then proceeded to prepare his bed. He pulled out a wooden plank and laid it across the front seat of his coupe so that part of the plank extended outside the open door on the driver's side of his coupe. He then stretched out and fell asleep:

The bottom half of the steering wheel was sawed off so that he could turn without waking himself up.

Dibblee was only half of the equation that would lead to the discovery of a major oil reservoir for Richfield executives—and that would lead to uncovering compelling evidence for hundreds of miles of movement along the San Andreas Fault. The other half was similarly quiet and highly capable young field geologist Mason Lowell Hill.

Hill was also a Californian, born in Pomona, east of Los Angeles and about 20 miles south of the San Andreas Fault. He attended the University of California at Berkeley for one year, where, according to Hill, he "suffered" through one of Lawson's classes. He thought Lawson was "opinionated and intolerant," though Hill admitted that Lawson was "a highly competitive and competent geologist."

After his single year at Berkeley, Hill enrolled at the University of Wisconsin, where he became fascinated—and obsessed—with earthquake faults, focusing on how to recognize them in the field and how they worked.

Hill was hired by Richfield Oil Company to join Dibblee and a half dozen other young geologists to search for oil in the Coast Ranges between Bakersfield and Santa Barbara. As was common practice, Richfield Oil employed a few local ranchers as "scouts" to keep company executives informed in case anyone started to drill a well and to send a message immediately if something important was found. On New Year's Eve, 1947, one of the "scouts" sent a message to a Richfield executive.

A few months earlier, the Norris Oil Company, an independent operator and one of Richfield's rivals, had commenced drilling a well close to an oil seepage on the side of Chalk Mountain at the northern end of Cuyama Valley. Norris, so the scout told the

Richfield executive, had hit "an oily sand" at a depth of a few thousand feet.

That set the Richfield Oil Company in motion. The executives were all in Pasadena, planning to attend the Rose Bowl game, when the call came from the scout. They abandoned their plans, and by driving cars and flying in chartered airplanes, they fanned out across the country, making personal contact with the owners of the various land plots in Cuyama Valley (most were absentee landlords who leased their lands to local ranchers). Within two weeks, to the shock of their competitors, Richfield had acquired oil leases to over 150,000 acres of land, almost the entire valley. The executives then sent Dibblee and Hill to decide exactly where to drill.

For the next few weeks the two men studied the area, Hill recalling that he did "some reconnaissance work" while Dibblee "did some semi-detailed geologic mapping of quite a big area." Finally, by early February they had decided—though they could never explain how they made their decision, which is part of the continuing mystery of exploring for oil—where their company should drill: four miles south of the well drilled by the Norris Oil Company, on the Russell Ranch at the north end of Cuyama Valley.

Four tense months followed. Executives at Richfield had bet the company's financial resources—and its future—on the recommendation of two young unorthodox geologists. On June 13, 1948, several executives were at the drill site evaluating the progress. They were with the ranch owner, Hubbard "Hub" Russell, a barrel of a man who said early that day he had his doubts whether anything dramatic would ever happen. What followed was a scene reminiscent of a Hollywood movie.

By then, the oil well had reached 2,800 feet below the ground surface and drillers were trying to punch through a hardened layer of sandstone. When they did, on that day the well quickly sank through a thick layer of loose oily sand. Those who were there heard the liquid crude come rumbling up, under pressure, and watched as liquid oil gushed out of the top of the well like a giant fountain. One of the executives who was present, a Mr. Charles Jones, would later tell others he had never been sprayed by a "finer cologne."

Hubbard Russell, who was standing next to Jones when the oil gushed, asked, "What does it mean?"

Jones responded, "It means you are a very rich man."

And the Richfield Oil Company was rich as well. With this discovery, the company more than doubled its oil reserves. Moreover, while most oil wells in California produce at the rate of 100 or so barrels a day, this first well in the Russell Ranch oil field at its peak would produce more than 500 barrels a day. Also it was what was called "sweet" crude, which means it had a low-sulfur content and would be easy and inexpensive to refine.

On more than one occasion I have stood along State Highway 166, which runs along the edge of the Russell Ranch oil field, and wondered why Dibblee and Hill chose this spot. By their own admission, there was nothing specific that attracted their attention. On the surface is a layer of boulders and sand washed down from the Caliente Mountains to the east. The field seems to be beneath a featureless plain. And yet later work would show that they hit the exact center of the oil field, a field five miles long and barely a half mile wide. If they had selected a spot just a quarter mile to the west or to the east, the exploratory well would have missed the reservoir. So it might have been a lucky hit, but less than a year later Dibblee and Hill did it again, this time along the south edge of Cuyama Valley, where a larger and more productive field was found.

For the next few years, Dibblee and Hill continued to search for oil in the Coast Ranges, extending their work eastward from Cuyama Valley into Carrizo Plain and closer to the San Andreas Fault. No more oil reservoirs were found.

A few years after the discovery of oil beneath the Russell Ranch—the layer that contains the oil is now locally known by oilmen as the "Dibblee Sand"—Thomas Dibblee left Richfield Oil Company and accepted a position as a geologist working for the federal government, extending his study of California's geology into the Mojave Desert. His first assignment—this was the early 1950s—was to search for borate deposits that could be used to make solid rocket fuel. He also continued to focus his attention on the San Andreas Fault.

In all, Dibblee's career as a field geologist lasted 68 years. During that time, he prepared more than 400 detailed geologic maps of California that cover more than 40,000 square miles—more than half of the state. It was a phenomenal achievement, and some of his maps, including the one that shows the San Andreas Fault running through Carrizo Plain, are regarded by some of his admirers as works of "art." I have seen that particular map framed and hung in private homes.

He never explained—and no one has ever duplicated—his ability to use his eyes to decipher quickly the geology that lay beneath his feet. But this much is sure: Thomas Dibblee, the skinny and somewhat awkward man who could be seen roaming California's landscapes, and who seemed to have a less than industrious approach to field geology, managed to cover and to understand more of California's geology than anyone ever had or probably ever will.

Hill continued to work for Richfield. In 1957, he was part of the team of geologists who found the first oil reserves in Alaska, along the Swanson River on the Kenai Peninsula, a discovery that helped persuade Congress to admit Alaska two years later as the 50th state. In the 1960s, he led the team that discovered oil along the North Slope at Prudhoe Bay in Alaska, the largest oil field yet discovered in North America. In 1966, Hill's company, the Richfield Oil Company, merged with the East Coast–based Atlantic Refining Company to form ARCO. And in 2000, ARCO was acquired by British Petroleum, today known as BP.

Regardless of their achievements in the oil industry, late in their lives when both men were asked what they regarded as their most important scientific accomplishment, they gave the same answer: It was providing a new and radically different perspective of the San Andreas Fault.

As late as the 1950s, anyone who seriously suggested the San Andreas Fault had more than a mile of accumulated horizontal movement—Levi Noble's work, by then, had been largely dismissed and ignored—exposed oneself to ridicule and derision. As a case in point, Don Wilhelms, who was then a geology student at the university at Berkeley—and who would go on to be one of the pioneers of the not-yet-established field of astrogeology, producing some of the first geologic maps of the moon—recalled how one of his professors, Nicholas Taliaferro, "didn't believe that the San Andreas had any strike slip," that is, any geologically important horizontal movement. Furthermore, as Wilhelms remembered, to make such a suggestion in front of Taliaferro was to subject oneself to a measure of loathing and, in some cases, to being called "a jerk."

Despite the name-calling, Taliaferro, who was a first-rate field geologist though apparently deficient in the art of effective rhetoric and in inspiring students, was voicing a long-held tenet in geology: The net effect of many earthquakes was to either raise or lower the land. It was simply impossible for huge rock masses to slide sideways any appreciable amount. Or so the thinking of the time maintained. Yet Hill and Dibblee knew that Taliaferro and those who held his opinion—and that was the vast majority of geologists—were wrong.

Hill and Dibblee, as well as Taliaferro, knew of Noble's claim that he had evidence for 25 miles of horizontal movement near Valyermo, but as early as the 1930s even Noble was beginning to question whether the Mormon Rocks and the seemingly identical rocks of the Punchbowl Formation had once been a single and continuous geologic unit, spliced and slid apart by the San Andreas Fault. Instead, Hill's initial attention was attracted to the long-forgotten work done decades earlier by a Berkeley student called Frederick Vickery.

In the 1920s, Vickery had studied a segment of the Calaveras Fault, one of the faults that splays off the San Andreas Fault south of San Francisco and runs into the mountains east of San Jose. Here, near Dublin, Vickery found an exposed section of sandstones and fossil-rich beds on the west side of the Calaveras Fault. Twelve miles to the south, on the *east* side of the Calaveras, he found an identical section, complete with the same diagnostic fossils. To Vickery, the conclusion was obvious: The two exposed sections of sandstones and fossil-rich beds had been sliced through by the Calaveras Fault, then after a series of major earthquakes slid to their present positions. Moreover, as Vickery mentioned—and Hill certainly noted—the direction of sliding along the Calaveras Fault was in the same direction as for the San Andreas Fault in 1906: offset to the right. But Vickery's conclusion did not sway Taliaferro or apparently anyone else because it was ignored as soon as it was published in 1925 and, according to Hill, "received no reaction." But it did influence him.

Of greater influence to Hill was a pair of scientific papers, both published in the 1920s, which Hill first read in the 1940s as he was searching for areas that might contain oil, which described two geologic sections on opposite sides of the San Andreas Fault: One was in the Gabilan Range near San Juan Bautista, close to the southern end of the 1906 rupture, and the other was far to the south in the San Emigdio Mountains north of Los Angeles. In reading these two papers, Hill realized that the two sections were identical, that they had probably once been the same geologic unit. In both cases, at the bottom of each section were layers of siltstones and sandstones. And in both cases, above the siltstones and sandstones was a red bed of the Miocene Epoch that contained a distinctive type of fossilized clam—*Mactromeris rushi.*

Above each of the red beds, whether at San Juan Bautista or in the San Emigdio Mountains, were dacitic and andesitic volcanic rocks. The reason these two papers that described these two geologic sections attracted Hill's attention was that the Gabilan Range near San Juan Bautista and the San Emigdio Mountains lie on *opposite* sides of the San Andreas. Furthermore—and this is the remarkable part—these two identical geologic sections, which locally are only a few miles in extent, are *175 miles apart*! That is comparable to the distance between San Francisco and Fresno, or between Fresno and Los Angeles.

But one or two examples are not enough evidence to overturn a widely held belief, so Hill knew he had to find additional ones. For that, he turned to Dibblee and his detailed maps.

To draw a geologic map in the days before aerial imaging—either by aircraft or by satellite—can be likened to being a flea that tries to comprehend an elephant. Never can the totality be seen, yet using empirical knowledge, such as Steno's laws of superposition,* the expectation of facies patterns,** and other droll principles, and employing the use of various colors, patterns, and line weights as well as a host of hieroglyphic symbols, it is possible to depict, in abstract form, the pattern of rocks that lie beneath one's feet.

The deciphering of a geologic map also requires special skills such as the ability to smooth out folds and slide rock masses along faults and to perceive how the rock units originated and how they have evolved. In that regard, Hill and Dibblee were a complementary pair: Dibblee could produce the maps as he roamed the countryside, and Hill could decipher them and understand the long geologic history of a region and put it into something the layperson (especially the oil executives) could understand. When Hill mentioned his idea about the fossils in the Gabilan and San Emigdio Mountains and whether they indicated that these two widely separated geologic units might in fact have once been connected as a single unit, Dibblee gave his usual noncommittal answer: "No reason why it shouldn't be." Thus, these two men came to understand how the

---

* Steno's law: Layers of rock are arranged in a time sequence with the oldest at the bottom and the youngest at the top.

** Expectation of facies patterns: The facies, or physical characteristics of a sedimentary rock, will vary within a sedimentary layer because the depositional environment varied; for example, coarse rocks are deposited near a river's source where water flows fast, while sand and silt are deposited near the mouth of a river where water flows slowly.

San Andreas Fault actually worked—how it moved and shaped the landscape around them.

In all, from Dibblee's maps and from reports published in scientific journals, they identified a dozen pairs of rock units that had once been single units but had been divided and displaced by the San Andreas Fault. These units ranged from monotonous gravels in the Santa Cruz, Temblor, and San Gabriel Mountains to giant blocks of granite found all along the fault.

Hill announced their findings in 1952 at an annual meeting of the Pacific Section of the American Association of Petroleum Geologists. The meeting was held at the Statler Hotel on Wilshire Boulevard, then the grandest and one of the newest hotels in Los Angeles with more than 1,000 spacious rooms, each one, so it was boasted, equipped with a new amenity—a television set.

Hill gave his presentation on the morning of October 30, the third presenter of the day. He explained how rock units on opposite sides of the fault could be correlated *if* they were slid long distances *horizontally* along the San Andreas Fault. Moreover, he showed that the *older* the paired rock units, the *greater* the amount of offset: Gravels that had formed 1,000,000 years ago had been offset tens of miles, while blocks of granite, which had been sliding for tens of millions of years, were offset more than 300 miles.

More than 1,000 people attended Hill's presentation. One was Henry Walrond, a recent college graduate who was beginning a career as an oil geologist. "My response to Hill's talk was not arbitrary or based on youthful brashness," Walrond would remember, "but was a reaction to a proposal that conflicted with my geologic experience." And that seemed to be true of everyone in the room.

Walrond remembered that after Hill's presentation, the mediator took the unusual step of asking for a show of hands of those who had been impressed by Hill's presentation. Most of those in the audience responded favorably. Walrond did not raise his hand. Then the mediator asked who actually agreed with Hill's conclusion. No one raised a hand.

Even Hill, evaluating the reaction years later, would say that his presentation had "shocked" the audience, and so he wasn't surprised that no one immediately rushed to agree with him so publicly. It also generated interest and controversy and caused others to reevaluate their work.

More than 60 years have now passed, and the Statler Hotel is gone. But the evidence introduced by Hill and Dibblee and their proposal that there is geologic evidence for tens to hundreds of miles of horizontal slip along the San Andreas Fault have been confirmed at scores of sites. Most sites are difficult for a nongeologist to visit and examine and be convinced that, indeed, the land has slid by large amounts. But there are a few notable ones.

The place to start is the first example Hill gave in his presentation—the basin immediately east of Cuyama Valley where he and Dibblee made their oil discoveries: the Carrizo Plain.

In the Coast Ranges, midway between San Francisco and Los Angeles, is a desolate, semi-arid, landlocked basin that is favored by birdwatchers and botanists. Migratory birds stop here semiannually. The basin contains the longest stretch of native grass that still exists in California. For those interested in prehistoric cultures, there is a remarkable concentration of multicolored paintings done hundreds to thousands of years ago by native peoples who lived, hunted, and traded in the area. And for those who seek evidence of Earth's episodic internal forces, few places offer a clearer example than Carrizo Plain.

For someone making a first trip to Carrizo Plain, I advise climbing a central hill, Overlook Hill, to get a view of the entire expanse. To the west is the Caliente Range, and beyond it is Cuyama Valley. Turning to the east, the first major feature is Soda Lake, its harsh white surface blinding to the eyes. And beyond Soda Lake, forming the horizon is a second range of mountains, the Temblor Range.

Temblor is the Spanish word for "shaking," and so one should not be surprised that the San Andreas Fault runs along the base of the Temblor Range. From the top of Overlook Hill, one can see about 40 miles of the fault trace, or about one-twentieth of the total length. It is best to climb Overlook Hill in the early morning or late afternoon when the sun is low in the sky and the shadows are long. At such times, the eyes are guided from the snow white of Soda Lake across a large grassy patch to a row of barren hummocky hills that mark the western edge of the Temblor Range. The edge of those hills is remarkably straight and is the San Andreas Fault.

To reach the fault from Overlook Hill, one drives southeast several miles, then backtracks to the northwest a similar amount, skirting the southern edge of Soda Lake. Take note of the occasional giant kangaroo rat, *Dipodomys ingens,* that seems to pop up out of nowhere and race across the road. The road may be barred temporarily by rolling tumbleweeds. At Elkhorn Road turn right, and after another mile arrive at Wallace Creek.

Before seeing the main attraction, I suggest walking a quarter mile to the base of the hills. Stand facing the southeast. To the left is a steep slope. To the right is the edge of a nearly level plain. Where the base of the slope meets the level plain is the San Andreas Fault. Now start walking.

No matter where you begin, eventually off to the right you will find a small dry streambed. With your eyes, follow the course of the streambed eastward to where it ends abruptly at the base of the slope. This is a beheaded stream. Where the stream is beheaded is the San Andreas Fault.

Next, walk a half dozen or so steps to the southeast, then look left. More often than not, on the adjacent slope and continuing up the hill is a straight and deep gully. Try it again. Find another beheaded stream, trace its course to the base of the slope, march off a half dozen or so paces to the southeast, look to the left, and voilà, there's another gully! You can do this over and over. What has happened here?

Each streambed-gully pair was once a continuous feature, formed by outbursts of rainwater from occasional heavy storms that left deep erosional scars that ran straight down the hillside and out onto the plain. The outwash and erosion happened repeatedly. Then in 1857, an earthquake—the Fort Tejon earthquake, the largest so far in California's brief history—shifted the ground *horizontally* at Carrizo Plain by about 20 feet. But this was not the only earthquake to occur along the San Andreas Fault at Carrizo Plain. To see what multiple earthquakes sliding along the same segment of the San Andreas Fault can do, walk back to Wallace Creek.

Standing on the hillside, about 40 feet above the plain, one sees a deep gully—at this scale properly called an *arroyo*, again, a Spanish word—about 100 feet across and 30 feet deep. With your eyes, follow its straight course down the Temblor Range as far as the San Andreas Fault. At that point, the creek makes an abrupt turn to the right, then after another straight course, this one 400 feet, the arroyo turns abruptly to the left and opens onto the plain. Geologists have known of this peculiar feature since 1909, when they were led to it by local ranchers, but its exact relationship to the San Andreas was not understood until the late 1940s when Robert Wallace—the namesake of Wallace Creek—began a lifelong interest in the offset streams.

Wallace began his geology career as a student mapping a 20-mile segment of the San Andreas Fault near Palmdale—the segment immediately west of where Noble worked 20 years earlier in the 1920s. Wallace was doing his initial geology work at the same time that Hill and Dibblee were successfully finding oil in Cuyama Valley. Being a student, he had almost no money, so he slept under the stars in the desert, listening to coyotes howl, urging them on with a violin he played, and surviving on cans of Campbell beef soup that he bought from the sole store in Palmdale. He chose to eat the soup cold.

It was along that segment of the San Andreas Fault, near Palmdale, that Wallace realized that many stream courses were abruptly offset at exactly the place where each stream crossed the fault—and always in the same direction: If one stood on one side of the fault, the other side had moved to the right.

That simple observation naturally drew his attention to Carrizo Plain, where eventually he documented dozens of stream offsets—again, always to the right—speculating that hundreds of examples probably exist along the entire fault trace from Humboldt County to the Salton Sea. Later investigators found those other offset streams, hundreds of miles away. They are in the Mecca Hills near Indio close to the Salton Sea, along the popular Sawyer Camp Trail that runs along the center of San Andreas Valley, and a right step in the source of Alder Creek just before it enters the ocean near Point Arena far north of San Francisco. An offset has even been found at the head of Noyo Canyon on the floor of the Pacific Ocean where the San Andreas runs under the ocean between Point Arena and Shelter Cove near Eureka. So Wallace's original estimate of hundreds of stream offsets should not seem that outrageous. They are the most obvious clue—at least to the nongeologist—as to how the fault works. And the most accessible and clearest example—and the one mentioned most often and illustrated in textbooks—is Wallace Creek itself.

To see a more subtle—though, when one realizes what one is seeing, a more startlingly—example of large horizontal movements along the fault, from Wallace Creek drive southeast on Elkhorn Road to the intersection with Panorama Road. Now start walking downhill to the west, paying attention to the rocks lying on the ground surface.

At first, the surface is covered with sand. Eventually, boulders are seen scattered across the surface. Examine them. Some are granite, a hard rock with visible crystals of white quartz and plagioclase, pink feldspars and black biotite and hornblende. Where did the granite come from?

To the east, behind you, are the rocks of the Temblor Range, mostly white shale. To the west, 10 miles away across Carrizo Plain, is the Caliente Range, which is mostly marine sediments. So where *did* the granite come from?

Now look to the south. You can see a flat-topped mountain. That is San Emigdio Mountain and it is composed of granite. And it is on the *east* side of the San Andreas Fault. When you walked away from the intersection of Elkhorn and Panorama Roads and encountered the boulders of granite, you were on the *west* side of the San Andreas Fault.

San Emigdio Mountain is 50 miles away.

A simple calculation, assuming the ground shifted horizontally 20 feet during every major earthquake, shows that about *10,000* earthquakes—each one comparable to the 1857 event—could have slid the granitic boulders 50 miles from their source at San Emigdio Mountain to Carrizo Plain. A remarkable result, but it is only the beginning.

A hundred miles north of Carrizo Plain is Pinnacles National Monument, a place favored by hikers and climbers because of the many rocky spires, crags, and other points of sharp relief, all the product of the deep weathering and erosion of an ancient volcano. Pinnacles National Monument is also a prime place to see the California condor (though don't make the mistake I made on my first visit and confuse the much more common red-headed turkey vulture, which also soars high on air currents, for the more majestic condor).

A geologic study conducted in the early 1970s identified ten distinct layers of volcanic rocks at Pinnacles, including successive layers of pumiceous tuff, hypocrystalline hypersthene-andesite, and augite-olivine andesite. Based on radiometric dating, the age of the volcanic rocks at Pinnacles—that is, the time of their eruption—is 23 million years. Pinnacles is immediately *west* of the San Andreas Fault.

Far south of Carrizo Plain are the rocks of the Neenach Volcanic Formation, which has ten distinct rock layers that match those at Pinnacles, including successive layers of pumiceous tuff, hypocrystalline hypersthene-andesite, and augite-olivine andesite. Radiometric dating of Neenach shows that these rocks also erupted 23 million years ago. By every measure, the rock sequences at Pinnacles and at Neenach are identical. And the Neenach Volcanic Formation lies immediately *east* of the San Andreas Fault.

Clearly, the Pinnacles-Neenach pair was once part of a single volcano, now sliced through and slid apart by the San Andreas Fault. This separated pair provides compelling evidence that the accumulated slip along the fault exceeds 100 miles—Pinnacles and Neenach are 175 miles apart.

But the evidence is not easy for a nonexpert to see. In fact, it is difficult even for a professional geologist to see, requiring extensive fieldwork, chemical and mineralogical work done in a laboratory and, in the case of Neenach, access to private land. For these reasons, when people ask me where definitive proof can be seen for more than 100 miles of movement along the San Andreas Fault, I direct them elsewhere: I send them to another pair of identical rock outcrops located farther to the south.

A few miles east of where the I-5 freeway passes through the San Gabriel Mountains north of Los Angeles—east of the steep downgrade known to truckers as "The Grapevine"—the San Andreas Fault cuts through the edge of Liebre Mountain. This mountain mass is difficult to distinguish from the other surrounding and equally rugged mountain masses except for a key feature: Within Liebre Mountain is a Triassic monzogranite that can be reached by driving along a barely maintained paved road, the Old Ridge Route Road.

A monzogranite is a type of granite that is the last part of a magma body to solidify. For that reason, it often has unusual mineral and chemical compositions and especially large crystals.

The one at Liebre Mountain—which solidified 215 million years ago during the Triassic Period and was reheated 70 million years ago during the late Cretaceous—catches the eye because it contains large well-formed crystals—often more than an inch across—of a salmon-colored potassium feldspar.

From Liebre Mountain in the San Gabriel Mountains, drive 120 miles southeast, through Valyermo and past the Pelona schist in Cajon Pass, to the San Bernardino Mountains, which are on the opposite side of the San Andreas Fault from the San Gabriel Mountains. Follow State Highway 330 toward Big Bear ski resort. Just two miles south of Angel Camp is a high roadside cut where an equally distinctive and attractive Triassic monzogranite is exposed. It also has large salmon-colored feldspar crystals that solidified 215 million years ago and were reheated 70 million years ago during the late Cretaceous.

These two widely separate exposures of a Triassic monzogranite have been subjected to a battery of laboratory tests, and in every case they have been shown to be identical. They have the same mineral content. They have the same chemical composition, which includes an unusual enrichment of strontium. So just as Pinnacles and Neenach can be paired, so can the two Triassic monzogranites in the San Gabriel and San Bernardino Mountains, a clear indication of more than 100 miles of horizontal sliding.

Scores of other pairings have also been made. Deposits of similar rock types are found on the west side of the San Andreas Fault in the northern Galiban Range near San Juan Bautista and on the east side of the southern Temblor Range on the Carrizo Plain. A light-colored sandstone known as the Butano Sandstone, originally deposited as a submarine fan, now located in the Santa Cruz Mountains, has been matched with an identical sandstone in the northern Temblor Range, indicating *200 miles* of horizontal movement.

But how did such huge displacements—caused by the action of tens of thousands of major earthquakes over the course of millions of years—come about? Hill and Dibblee had no explanation. In fact, it would be another decade before the answer became clear. It would be provided by a Canadian geophysicist who had never seen the San Andreas Fault but who, fortuitously, had made a trip to the Hawaiian Islands.

CHAPTER 8

A TRANSFORMATIVE IDEA

The San Andreas Fault is here postulated
to be a dextral transform fault.
—J. Tuzo Wilson, 1965

In October 1958, a team of scientists from the National Academy of Sciences was planning to go to Antarctica to evaluate the current progress of research there. Merle Tuve, a member of that team, had a heart attack. So on short notice, J. Tuzo Wilson, another member of the academy, went in his place.

One might think Wilson was weary of travel, having completed more than 100,000 miles during the previous 18 months, but this was the end of the International Geophysical Year, and as president of the Union of Geodesy and Geophysics—an organization that was a primary supporter of the geophysical year—he was anxious to meet and organize and collaborate with scientists who had agreed to make simultaneous observations of the atmosphere, the oceans, and the land surface over the entire Earth. It was an unparalleled opportunity to see the planet as no one had ever seen it—from remote scientific stations on ice caps, in deserts, at the tops of mountains, and from the shifting decks of ships that were crisscrossing the oceans probing the deepest parts and, whenever the opportunity arose, seeking out and intentionally sailing into storms.

Wilson began his trip to Antarctica by leaving his home in Toronto, where he was a university professor, and flying to Chile, where he joined the other five members of the team. From there, this sextet of some of the world's leading scientists flew in a military transport plane bound for a six-week tour of scientific stations on the southern continent, a part of Operation Deep Freeze organized and led by the United States Navy. After those six weeks, the team returned north via New Zealand, then to Honolulu in the Hawaiian Islands, where they inspected a newly installed solar radio telescope near Diamond Head. After that, the team divided. Half went to Maui to see one of the new worldwide stations that had just begun to track the first orbiting artificial satellites. The other half, which included Wilson, made a longer trip to Hilo on the island of Hawaii, where a nearby weather station, equipped with the latest in meteorological instruments, was set to open.

It is a 40-mile drive from Hilo to the weather station, and Wilson recalled that they first passed through fields of sugar cane, then miles and miles of tropical forest that changed suddenly into a field of dark recently congealed lava. They bumped and jolted up and over the stark landscape, climbing ever so slowly higher, the air growing colder and the sun overhead more brilliant.

At 11,000 feet, with a few snowflakes in the air, they ended their journey in front of three metal huts. Wilson was surprised by the absolute stillness in the air. It was cold, and he and his two comrades went inside one of the huts, where they were greeted by three local scientists who ran the station and who immediately began to describe their work.

Today the station is known as the Mauna Loa Observatory. It would be here, just a year after Wilson's visit, that scientists would announce the first measurements that showed that the amount of carbon dioxide in the atmosphere was increasing at a high rate. But on the day he visited, December 4, 1958, that equipment was still being tested.

After the scientific discussions and after a lunch, the six men went outside. The wind was now blowing. Wilson walked off to look around. Only years later, reflecting on that visit, did he realize that he had arrived at a revolutionary way of understanding the Earth.

That afternoon he was standing high on the slope of Mauna Loa, the world's largest volcano and one of the most active. To the southeast, out of sight of the weather observatory, was another volcano called Kilauea, much smaller than Mauna Loa though more active. Directly to the north was Mauna Kea, another volcano, that one dormant. And beyond, to the northwest, stood Maui, rising from the sea and on Maui was another volcano, Haleakala, still active though less so than either Mauna Loa or Kilauea, and more deeply dissected by erosion than Mauna Kea and so probably older than that volcano.

And beyond Maui, though he could not see it, was a long chain of volcanic islands, including Oahu, where Honolulu is located. Each successive island in the chain was more deeply eroded—and, hence, must be older—as one progressed northwestward along the chain. Far beyond those islands was a line of islets and shoals, barely rising above sea level and familiar to fishermen, that were also volcanic and certainly older than the main islands.

While Wilson was thinking about this age progression of islands and islets and shoals out in the middle of the Pacific Ocean, he also thought about a discovery announced recently: It seemed that the entire seafloor of the northern Atlantic Ocean was moving, at an incredibly slow though steady pace, away from a long range of subdued mountains down the middle of the Atlantic Ocean and known, appropriately, as the mid-ocean ridge.

And so, Wilson thought, as he collected and ruminated on these fundamental observations over the next few years, might the seafloor of *every* ocean be in slow and steady motion? Might the entire floor of the Pacific Ocean, extending over a third of the planet, be in slow and steady motion? And if so, what would be the result? Could it explain the age progression of islands and islets and shoals?

He decided that it could, *if* the Earth's entire surface was sliding. From this simple realization would come the foundation of the theory of plate tectonics.

Born in Ottawa in 1908, Wilson began to use his middle name, "Tuzo," early in his professional career to distinguish himself from another geologist named J. T. Wilson, who worked at the University of Michigan. Tuzo was his mother's maiden name, originally Tuoselle or Touzelle, later corrupted to Tuzo, which came from Huguenot ancestors who had landed in Virginia in the 17th century.

Besides giving her son an unusual middle name—which makes it easy for science historians to identify him—his mother, Henrietta, or "Hettie," had another major influence. She was an accomplished mountaineer, having scaled, before her son's birth, several high peaks, first in the Alps, then in the Rockies, at a time when few women engaged in such activity. Her most notable feat was to be the first person to reach the summit of Peak No. 7 in the Valley of Ten Peaks rimming Moraine Lake in the Canadian Rockies. She made her ascent in 1906. The next year, in her honor, the 10,630-foot-high peak was named Mount Tuzo.*

Inspired by his mother, the son also became a mountaineer, being the first to make a solo ascent of Mount Hague in Montana. It was a desire for such strenuous outings that made him decide to be a geologist.

In 1930, Wilson graduated from the University of Toronto—the same university Andrew Lawson had graduated from nearly 50

* Because of their scenic qualities, Lake Moraine and Mount Tuzo were depicted, for many years, on the reverse side of the Canadian $20 bill.

years earlier—and accepted a job with the Canadian Geological Survey—again, as Lawson had after his graduation. Wilson's first job was to prepare a geologic map of southern Nova Scotia, an assignment that was filled with adventure. On one occasion, when he was in a canoe, he spotted a moose swimming in the water and pulled up to it. He jumped on the moose's back and rode it a short distance. Another time, when he and his camp mates were short of food, he again saw a moose while he was in a canoe, though this time he hit the animal on the head with an axe, killing it and providing his camp with fresh meat.

He worked for the Canadian Geological Survey for ten years. At one point, he was assigned the task of preparing a geologic map of all of southeastern Canada. That experience, he would say, conditioned his mind: Everywhere he worked and all the data he compiled indicated the Earth's surface was stable. Seldom did he encounter a fault or a fold. He saw that old rocks, such as those of the Mesozoic Age, lay atop even older rocks, those of the Paleozoic, that were on top of even older rocks of the Precambrian—and all lay in correct sequence with no sign of having been fractured or displaced. From that, Wilson concluded—and for years argued passionately for—the view that, by and large, the Earth's surface was immobile, that at the largest scale the present outlines and configurations of the continents and ocean basins had not changed in any significant way over the eons of Earth history. It was an idea, Wilson reminded others during debates, that had originated hundreds of years earlier by the most venerated of scientists, Isaac Newton, who suggested the entire Earth, including its outer rim, was rigid. And Wilson saw no geologic evidence to the contrary.

Wilson, of course—as did almost every geologist of the time—passionately discounted any theory to the contrary, which included the one proposed in 1912 by German meteorologist Alfred Wegener that the continents had shifted their positions a significant amount during geologic history, an idea that was later corrupted into the phrase "continental drift."

Wegener based his theory on a host of observations, including evidence of glaciation in tropical regions, the discovery of identical fossils at widely distant locations—such as *Glossopteris*, a giant fern-like plant with tongue-shaped leaves that could be found in Africa, South America, Australia, India, and Antarctica—and the existence of similar rock sequences on opposite sides of the Atlantic Ocean. Wilson and others thought these were misinterpretations of the geologic record. Also, Wegener could offer no physical mechanism to explain why the continents should move. As Wilson wrote in 1949, "Continental drift is without cause or a physical theory." Later, he would recant that statement and confess that at that time he had been "too stupid to accept, until I was fifty" the idea that the Earth's surface was, indeed, highly mobile. But that was years in the future. Throughout most of his career, he maintained that the Earth was rigid and strong, and that the surface did not slide any appreciable amount.

By the 1950s, now a professor at the University of Toronto and having become somewhat of a scientific gadfly, Wilson began to temper his stance for a rigid Earth and conceded that the cooling of a hot interior would have caused the planet to contract, giving rise to mountain ranges and island chains. By 1960 he modified his view again, this time arguing that the recent detailed mapping of the mid-ocean ridge—which was now known to extend beyond the north Atlantic Ocean as a long continuous line of mountains that snakes along the ocean bottom like the seams of a baseball—showed that this was the place where the Earth's surface was being pulled apart, driven by a general expansion of the entire plate. He suggested that the reason for the expansion—which was counter to his earlier claim that the planet was contracting—was either because the Earth's interior had a high concentration of radioactive elements and thus the interior was getting hotter, or because the force of gravity was decreasing. Neither explanation is true, so it was fortunate Wilson's pursuit of expansionism was short-lived.

Just a year later, in 1961, he dramatically changed his idea about the planet and its global dynamics yet again. By then, a host of new discoveries—many of them made during the International Geophysical Year*—had been announced and debated and confirmed. For example, it was now known that the deepest part of an ocean is not in the middle, but along the edges at narrow trenches; that the magnetic poles had apparently wandered greatly; and that ocean basins were much younger than continents. These seemingly disparate observations—and many others—forced Wilson to reconsider what he thought he knew about the Earth. The man whose groundbreaking insights caused Wilson's reevaluation was Harry Hess of Princeton University.

It is fair to say that Hess had a strange fixation on rocks, as his wife, Annette, could attest. On the only vacation they ever intentionally took away from geology, which was a honeymoon on Nantucket Island, she would remember the island as having "only one rock, and that was brought in as a monument." She also remembered that her new husband "used to look longingly at it."

Hess was a quiet, unpretentious man always sporting a small mustache and always toying with a lit cigarette. His office was one of legendary clutter, filled with bathymetric charts that he had assembled during scientific cruises. He had an uncanny ability to assimilate widely disparate observations into a theoretical whole, as he displayed in 1962, when he proposed a new idea about the origin of the seafloor, an idea so radical that instead of calling it "science," he referred to it as "geopoetry."

The idea was this: Material from the Earth's interior rose along the mid-ocean ridge, causing the ocean basins to spread apart. A case in point was the northern Atlantic Ocean, where Hess and

* An international scientific project that actually lasted 18 months, from July 1, 1957, to December 31, 1958, and that involved scientists from 67 countries who coordinated their research to understand the atmosphere, the oceans, and the solid earth. The first two artificial satellites were launched under this project, Sputnik by the Soviet Union and Explorer I by the United States.

others had conducted oceanographic cruises and shown that the entire Atlantic Ocean was spreading apart. The spreading explained why the coastal outline of Brazil in South America fits neatly against the west coast of Africa, a fact long noticed by cartographers but, until now, never satisfactorily explained. Overall, the seafloor moves away from the mid-ocean ridge like the top of a conveyor belt. And if seafloor is being created along a ridge, Hess reasoned, it must be consumed somewhere—and he pointed to deep ocean trenches, which he described as "jaw crushers," as the places where seafloor was slipping back into the Earth.

If true, it was a revolutionary idea. But how could it be proved that the seafloor was sliding slowly and steadily like the top of a conveyor belt? Wilson knew of a way.

He thought back to the afternoon he had stood on Mauna Loa. Conventional scientific wisdom held that the long chain of islands and shoals had formed simultaneously and the volcanic fires had ended sequentially from northwest to southeast, accounting for the current activity at the southeastern end at Mauna Loa and Kilauea. But if the seafloor moved, and if there was a *single* source of magma, then the chain of islands would not only show an age progression but new islands were still being created. Wilson explained it this way: Imagine lying on your back at the bottom of a shallow stream, blowing bubbles to the surface through a straw. The bursting bubbles are the Hawaiian Islands, and they lie in a line because they were swept along the surface of the moving stream.

The older volcanoes, he predicted, would lie farther from the source. And that was exactly what was revealed a decade later when radiometric age dating was applied to Hawaiian rocks, but that was still in the future. For his purpose, Wilson had to rely on a less direct argument.

He knew the Hawaiian Islands are not the only long chain of Pacific islands that show an apparent age progression. The Austral, Society, and Tuamotos also have active volcanoes at the southeast ends and the islands to the northwest appear to be progressively older.

He wrote this idea in a paper and submitted it to a leading scientific journal in the United States. It was rejected on the grounds that the idea of age progression in the origin of the Hawaiian Islands was at odds with everything else currently known about the geology of the major islands—that is, Kauai, Oahu, Molokai, Maui, and Hawaii. Undeterred, he sent it to a leading journal in Canada and it was published immediately.

Publication, however, does not insure interest, and his paper, "A Possible Origin of the Hawaiian Islands," was ignored, at least for now. Today it is regarded as one of the pioneering papers—one of the crucial steps—that led to a revolution in understanding internal Earth dynamics. It was the first time anyone had pointed to evidence, no matter how crude or circumstantial, that the Earth's surface was highly mobile, that a vast area—in this case the entire floor of the Pacific basin, one-third of the Earth's surface—was moving as a single, coherent, gigantic block over distances of many hundreds of miles.

But Wilson realized that a key element was still missing. If one looked at any of the new detailed bathymetric maps of the Pacific basin—such as the one prepared in 1963 by Gleb Udintsev of the Soviet Union's Institute of Oceanology, which shows how crucial international cooperation was in the 1960s in arriving at a theory of global dynamics despite the Cold War—one could see in the southeastern Pacific a mid-ocean ridge known as the East Pacific Rise, where, according to Hess, seafloor was being created, and along the west and north edges of the Pacific basin, running along the edge of the Mariana Islands, Japan, and the Aleutian Islands, deep trenches where the seafloor was being consumed. But what was happening elsewhere along the edge of the Pacific? Specifically, what was happening for hundreds of miles along the California coast where there was no mid-ocean ridge and no trench?

It was here, Wilson realized, that the final crucial piece to the global puzzle would be found. And to do so, he would have to explain the San Andreas Fault.

In the fall of 1964, Wilson arrived at Cambridge University in England to begin a one-year collaboration with British scientists who were trying to make sense of the huge volumes of data being accumulated about the seafloor. Little is known about the way he spent his first few months in England except that immediately after the new year, he went on vacation.

He hired a yacht and went sailing with his wife and two daughters on the south coast of Turkey. Whether he completely blanked his mind and relaxed on the trip or concentrated intently on the implications of seafloor spreading and the origin of Pacific islands is not known. But after a month, when he returned to Cambridge, he was brimming with a new idea.

John Dewey, a lecturer in structural geology at Cambridge, remembers Wilson bursting into his office at the Sedgwick Museum. "Dewey," Wilson exclaimed, "I have just discovered a new class of fault."

"Rubbish," Dewey responded. "We know about the geometry and kinematics of every kind of fault known to mankind."

Wilson then grinned and produced a simple model he had cut from colored paper. It consisted of two L-shaped pieces hooked together like a pair of hockey sticks. He slid the pieces back and forth so that the inside edge of one L-shaped piece slid against the inside edge of the other. He explained that where the two pieces of paper were sliding was a new type of fault, a *transform* fault, because it transformed the spreading movement of one mid-ocean ridge to the spreading movement of an adjacent, offset ridge.

"I was transfixed," Dewey remembered of the moment, "both by the realization that I was seeing something profoundly new and important, and by the fact that I was talking to a very clever and original man."

To put this in context, for the last few years scientists had looked at maps of the seafloor and seen the mid-ocean ridge that runs along the center of the Atlantic Ocean. The ridge actually consists of straight segments offset by faults, sometimes for hundreds of miles. The favored interpretation was that the faults were evidence that the entire ocean was being torn open, from coast to coast. But Wilson disagreed. The faults were evidence of tearing, but only between adjacent ridge segments. This new view suggested that the opening of an ocean basin was concentrated at the ridge and along the connecting faults—the transform faults. The rest of the seafloor simply drifted along as a rigid block until it was consumed at a deep trench.

Wilson gave a name to these rigid blocks; he called them "plates." Most of the floor of the Pacific Ocean, from the East Pacific Rise to the deep trenches, is a plate—the Pacific plate. From the eastern edge of the Pacific plate to the mid-ocean ridge in the North Atlantic Ocean is the North American plate. In all, Wilson suggested, there were a dozen or so major plates and many minor ones, each one sliding slowly across the Earth's surface. And the places where they met—the plate boundaries—were the most geologically active regions of the planet.

Suddenly, much became clear—at least to Wilson. If one returns to Mallet's map of 1857 that showed the major seismic zones of the Earth, these zones, which lie mostly along the edge of the Pacific and across southern Asia and into the Mediterranean, identify some of Wilson's plate boundaries. If one examines modern maps of worldwide seismicity, plate boundaries are defined in great detail. And those boundaries are of three types: the two suggested by Hess, spreading at mid-ocean ridges and the "jaw crushers" at deep trenches—the word "subduction" would not be used until 1969—and the transform faults suggested by Wilson.

But in 1965, while Wilson was still at Cambridge, the idea of moving plates was not yet accepted by most members of the geological community. So he needed a test.

After studying several maps, including the Udintsev map, which he had pasted to the wall of his office at Cambridge, he finally realized that the San Andreas must be a major transform fault system. He knew of Hill and Dibblee's claim of hundreds of miles of horizontal displacement along the fault—though once, when Wilson attended one of Hill's presentations, he had immediately dismissed the idea—and it was their claim that directed his attention to the San Andreas. More important, the south end of the fault seemed to connect with the East Pacific Rise, which oceanographers had traced as far north as the west coast of Mexico. But what was at the north end of the fault?

Hess happened to be visiting Cambridge in 1965, and he and Wilson, along with young British researcher Frederick Vine, were in Wilson's office one morning discussing the San Andreas Fault. Vine's specialty was the study of magnetic field anomalies on the seafloor, and he and others had shown a few years earlier that strong magnetic anomalies were always associated with the mid-ocean ridge.

"Look, there should be a ridge here," said Wilson, pointing to a map where the north end of the San Andreas Fault projected into the ocean.

"Well, if you're going to put a ridge there, then there ought to be some magnetic expression," responded Hess. Vine then hustled to the library to retrieve the appropriate map that would show the magnetic field.

Lo and behold, as Vine would recall, when they unrolled the map there it was: a linear magnetic anomaly where Wilson had indicated the crest of a mid-ocean ridge should be. Vine later admitted he stood there amazed. The map had been in the library for four years, yet no one had ever noticed so prominent a feature. Later, he and Wilson would give it a name—the Juan de Fuca Ridge, named after the nearby strait that separates Vancouver Island and Washington State.

That act of discovery elevated the San Andreas Fault from a feature of local interest to one of global importance. Ever since—as almost every geologic textbook now proclaims, and almost every speaker who has ever made a presentation about faults and global dynamics asserts—the San Andreas Fault has been known as a plate boundary, the line where the immense mass of the Pacific plate grinds against the equally formidable mass of the North American plate. And as a result, earthquakes must be a common occurrence in California, which they were.

But Wilson realized something else. If tectonic places could slide for hundreds or perhaps thousands of miles, then by necessity plate boundaries had to evolve. And if plate boundaries evolved, then a feature such as the San Andreas must have had a beginning. Soon after Wilson made his suggestion that the San Andreas Fault was a plate boundary, he and others engaged in finding an answer to this additional question—and they figured out how the fault had originated.

An abundance of paleontological, paleomagnetic, lithologic, and paleoglacial evidence—offered first by Wegener then by others, which led to the discovery of the mid-oceanic ridge, magnetic lineations, and additional fossil discoveries—points to the existence, a very long time ago, of a single large continental mass—a supercontinent—that scientists now refer to as Pangaea.

Inevitably, because the forces that drive the movement of plates today were in existence back then, Pangaea broke up. It first fractured along an east-west line and the two giant parts drifted apart, forming the Tethys Sea, a large ancestor of the Mediterranean Sea. Another fracture split Pangaea along a north-south line and the large fragments that drifted to the west became the two American continents while those that drifted east became Eurasia and Africa. If one wonders what the splitting and separation of the continents looked like during an early stage, one has only to examine the Red Sea, where, running along the seafloor, is a young mid-ocean ridge that is driving Africa and the Arabian Peninsula away from each other—the ridge defined, in part, by a line of persistent earthquake activity.

When Pangaea existed and while it was being broken up, the other side of the planet was covered by a vast ocean—the Panthalassa Ocean, which is Greek for "all seas"—which was a much larger version of today's Pacific Ocean. One of the things known for certain about the Panthalassa Ocean is that its floor was crossed by at least one mid-ocean ridge because segments of this prehistoric ridge still exist—but that is getting ahead of the story.

As the north-south split of Pangaea grew wider—forming a new ocean, the Atlantic Ocean, between the Americas and Eurasia and Africa—the two American continents drifted west, causing the Panthalassa Ocean to get smaller, its seafloor plunging beneath the American continents.

It is a process still seen today. Rocks on the seafloor—mainly basalt—are denser than continental rocks. Ocean basins are low and continental mountains are high because when a continent and an ocean basin collide, the denser rocks on the floor of the ocean plunge beneath the lighter rocks of the continent. The resulting plate boundary is a subduction zone with a deep trench. This is happening today along the entire west coast of South America, and for a long time it happened along the entire west coast of North America. But the situation changed when North America encountered the mid-ocean ridge that was already running along the bottom of the Panthalassa Ocean.

That mid-ocean ridge was, of course, the boundary between two plates. The plate moving away from the ridge to the west by northwest was and is still known as the Pacific plate. The other plate, the Farallon plate, named for the Farallon Islands west of the Golden Gate, was and still is moving to the southeast. North America got a head start over South America on its westerly movement—the north Atlantic opened before the south Atlantic—and that continent has drifted far enough west so that it has overridden a segment of the mid-ocean spreading ridge of the Panthalassa Ocean.

At the moment the North American plate touched the mid-ocean spreading ridge, the San Andreas Fault started to form.

To see it another way, the Juan de Fuca Ridge and the East Pacific Rise are segments of a once-continuous mid-ocean ridge of the Panthalassa Ocean—now the Pacific Ocean. What was once the Farallon plate is now, in the north, the Juan de Fuca plate, which moves eastward from the Juan de Fuca Ridge and plunges under Washington and Oregon. To the south, what still exists of the Farallon plate is now the Cocos plate off Central America and the Nazca plate off South America; both move eastward away from the East Pacific Ridge. And connecting the two segments of mid-ocean ridge—that is, the Juan de Fuca Ridge and the East Pacific Rise—according to Wilson, is the San Andreas transform fault.

Much detail has been added since 1965, when Wilson made his original proposal. The Juan de Fuca Ridge has been divided into two pieces, the northern one still known as the Juan de Fuca plate, and the southern one, off the coast of southern Oregon and northern California, is called the Gorda plate. Connecting the Gorda plate to the California coast is a long straight feature called the Mendocino escarpment. And where the escarpment touches the California coastline, near Cape Mendocino, that point is regarded as the northern end of the San Andreas Fault.

Features at the southern end of the fault have also been greatly refined. The East Pacific Rise runs north continuously as far as the opening to the Gulf of California where, from that point north, running along the axis of the gulf, there is a series of short spreading centers connected by short transform faults that continue north until they run across land as far as the southern part of the Salton Sea. Formally, the northernmost spreading center ends on the east coast of the Salton Sea at Bombay Beach, which is regarded as the southern end of the San Andreas Fault.

All this raises the question: *When* did the transition occur from subduction zone to transform fault? When did the earliest vestige of something like the San Andreas Fault form in California? To answer that question, one has to look to the sea.

Tanya Atwater was still in high school when Wilson was standing on the slope of Mauna Loa, thinking the Earth's surface was immobile. She was in college when he proposed the idea of mobile rigid plates, and had completed her degree in geophysics the year Wilson proposed the San Andreas was a transform fault—an indication of how quickly ideas were changing. Anxious to be part of the revolution in understanding how the Earth worked, Atwater attended the Scripps Institute of Oceanography near San Diego, where she hoped to do her own research that would reveal even more tantalizing details of the Earth's mobility. But initially she was stymied in her attempt—because she was a woman.

To make the crucial measurements that would change our view of the solid Earth and lead to the concept of plate tectonics—the phrase "plate tectonics" would not be used until 1969—one had to go to sea, but in the 1960s, to go to sea on an oceanographic ship, such as those operated by Scripps, one had to be a man.

"She would have no privacy," Atwater remembered was a common reason to exclude her. Another was that a woman would have no convenient or comfortable place to bathe or to urinate. And there was the possibility that just the presence of a woman working in the deep confines of a ship might be enough to tempt even the most self-disciplined male crew member.

Atwater also endured personal slights. She would remember that, when visiting other institutions, anxious to discuss her own work, she was often not introduced, others assuming she must be some guy's tag-along girlfriend.

But Atwater was fortunate in that her male mentors—at the time all senior scientists at Scripps were male—gave her the respect she deserved and insisted that she have the opportunity to work on a ship and go to sea. And when she went to sea, one of the things she noticed on those oceanographic cruises was that the pattern of seafloor magnetic anomalies off the California coast was different from those elsewhere in the ocean.

Here it is instructive to step back a bit. During the informal meeting held in Wilson's office when Vine unrolled the map of seafloor anomalies in the northeast Pacific, Wilson, Vine, and Hess all instantly recognized a pronounced linear anomaly that indicated the crest of the subsequently named Juan de Fuca Ridge. They also noticed something else. As Vine would recall the scene, the three men stared in amazement at the map because not only were there other linear anomalies parallel to Wilson's proposed ridge, but a symmetry to the pattern of anomalies about the ridge crest. And Vine soon had an explanation.

As Hess had suggested years earlier, hot material from the mantle rose along a ridge crest and formed new seafloor. Then, as Vine now explained, as the material cooled its temperature passed below the Curie point, the temperature below which a material with suitable minerals, known as ferromagnetic minerals, would become magnetized in the Earth's field. Mantle material had such minerals, so after the material cooled, it left a pronounced magnetic anomaly along the ridge crest. Then as the ridge continued to spread open, the now-cooled material moved off to the sides of the ridge and new, hot mantle material rose to replace it.

However, the direction of the Earth's magnetic field flips every hundred thousand or few million years, so that the north magnetic pole becomes the south magnetic pole and the south magnetic pole becomes the north magnetic pole. Why the magnetic poles flip is still not known, but the fact that they do produces an interesting pattern of linear magnetic anomalies on the seafloor—known as magnetic remanence—one that is often compared to zebra stripes.

All seafloor material that was at a ridge crest and cooled when the Earth's magnetic field was in its current alignment has the same direction of magnetization and is known as "normal."

Material that cooled when the Earth's field was in the opposite alignment is "reverse." And so, as a continuous supply of hot material rises from the mantle, cools, and moves off and away from a ridge crest—as Hess described it, in the manner of "a conveyor belt"—an alternating-line pattern of normal and reverse magnetic anomalies is produced. Therefore—and this is the important part for what Atwater would discover—if one finds the same pattern of normal and reverse magnetic anomalies on different sections of the seafloor, then those two sections of seafloor formed over the same age range. Furthermore, one can match individual anomalies on different sections of seafloor and know they were created at the same time.

The basic idea is straightforward, but is a challenge to put into practice. For example, the age of seafloor magnetic anomalies had to be determined from both radiometric age dating of lava flows on land and from the recovery of sedimentary samples taken from the seafloor that contained microscopic fossils. Also, it had to be determined whether seafloor-spreading rates were constant or ebbed and flowed.

It might seem hopelessly complicated, but Atwater found a way to do it by going on oceanographic cruises and collecting and compiling an immense amount of data, and from that making a detailed map of magnetic stripes across much of the northeast Pacific.

Her map, published in 1970, showed 32 stripes, each one identified with a "magnetic isochron number," or "chron." Chron 1 was the magnetic stripe currently forming over the crest of the Juan de Fuca Ridge. Chron 32 was found out in the Pacific almost as far as the Hawaiian Islands and she identified it as having formed along the crest of the mid-ocean ridge 75 million years ago.

All mid-ocean ridges spread symmetrically. But Atwater realized that the pattern of magnetic stripes off the coast of California has no symmetrical counterpart because the mid-ocean ridge that formed those stripes has disappeared—that is, it has slid under the North American plate. And that was the key: The age of the magnetic anomaly next to the coast of California would be the age of the first San Andreas–type fault, and thus how old the fault was would finally be revealed. After considerable work, Atwater identified the magnetic anomaly, Chron 8, and determined the age of the San Andreas Fault: 25 million years old.

The fault may be 25 million years old, but as far as human understanding of the fault and geophysics in general, it is astounding how quickly ideas progressed from Hess's geopoetry to Wilson's mobile plates and to understanding the origin of the San Andreas Fault, then to Atwater's success in answering a question that, a decade earlier, no one could have imagined to ask or, if someone then had such a fanciful notion, no one could have conceived how the question might be answered. And even more astounding considering how little men like Richter, Lawson, and others understood about the geological anomalies they were seeing and earthquakes they were feeling and measuring just a generation before. And it was just the beginning. Now a whole range of geologic puzzles could be solved.

One that had been perplexing geologists since the time of Lawson and his discovery of the San Andreas Fault was the origin of a chaotic mixture of mangled sandstones, shales, cherts, and basalts that comprise most of the Coast Ranges of California. That includes every major hill of San Francisco, including Telegraph Hill, Russian Hill, and Nob Hill, as well as Alcatraz Island in San Francisco Bay, and Twin Peaks and San Bruno Mountain south of the city, a complex assemblage that Lawson had named in 1892 the Franciscan Series.

What was puzzling and intriguing about the Franciscan Series was that it was impossible to put the various rock units in stratigraphic order. For decades, Lawson and, later, other geologists strained to use the tried-and-true methods of superposition—whereby rock layers are deposited in a time sequence with the oldest at the bottom and the youngest at top—to decide in what order the various sandstones, shales, cherts, and so forth had been deposited. But this conflicted with what fossils recovered from those same layers indicated about the relative ages. Fossils indicated that some layers higher in the sequence were actually *older* than layers underneath. The solution became clear after Wilson's work and the development of the theory of plate tectonics, after it was realized that ocean crust could subduct beneath continental crust.

To see the evidence for oneself, I suggest going to the rocky Marin Headlands immediately north of the Golden Gate Bridge. After gazing at the fantastic view of San Francisco's cityscape, turn around and look at the rocks exposed in the roadcut.

What catches the eye are the parallel ribbons of red chert folded into giant kinks and the occasional chevron. If the chert is examined closely—a glass hand lens is of great utility—one sees countless white dots. Those dots are skeletons of a tiny sea creature, a radiolarian, that built its tiny shell out of the mineral silica. Radiolarians thrive only in warm equatorial waters, and when they die their skeletons sink slowly to the ocean floor to form what is inelegantly though accurately known as radiolarian ooze. But how did the skeletal fossils of these tiny equatorial creatures get to California, and why can they be found interspersed among the various sandstones, shales, and basalts of the Franciscan? The answer: because they rode in, conveyor-belt style, on the Farallon oceanic plate, then were smashed against the continent of North America.

To elaborate, as the North American plate drifted west, the Farallon plate moved southeast. The collision formed a subduction zone as the denser Farallon plate slid under the lighter continental rocks of North America. As it did so, the brow of North America acted like a bulldozer and scraped off some of the rocks of the seafloor, forming what plate tectonophysicists call "an accretionary wedge." The rocks of this wedge are the crushed and mangled and highly deformed rocks of the Coast Ranges—which includes the red ribbon chert north of the Golden Gate—as well as the hills of San

Francisco, most of the mountain mass of Big Sur and its spectacular sea cliffs, and much more. In short, the Franciscan contains rocks that, quite literally, came from a wide stretch of the seafloor of the ancient Pacific Ocean, then were piled up against the northern and central coasts of California.

To see the full suite of rocks of the Franciscan, one could drive hundreds of miles along back roads that wind through the Coast Ranges. But there is a simpler way.

At a few places along the coast, hillside streams and ocean currents have combined to concentrate rocks washed down from the Franciscan at several beaches. The most famous is Moonstone Beach near San Simeon.

Here the beach is covered with pebbles of a wide variety of color, texture, and luster, making it a favorite of rock collectors—which is why one sees, at low tide, people walking slowly up and down the beach, their eyes fixed to the ground, their bodies occasionally bent down to examine a particular pebble. Among the most common of the pebbles are those of the red chert. Another is a drab gray sample of seafloor basalt. Pebbles of green serpentinite struck through with tiny veins of white chrysotile—that is, asbestos—are not uncommon. But the one that most collectors seek—and is the mistaken namesake of the beach—is known to locals as moonstone.

True moonstone is a low-quality gem that displays an iridescent, silvery moonlight-like quality. But there are no moonstone gems at Moonstone Beach. What people find is a white translucent stone, often with gray or brown bands, composed of the mineral chalcedony, a form of silica. Ironically, considering how some geologists flock to Moonstone Beach to examine pebbles of the Franciscan it is probably the best place to hold samples of California's diverse geology in one's hand—the pebbles of chalcedony that most rock collectors seek are not products of the collision of the Farallon and North American plates. Instead, they formed long after the seafloor was plowed up and the Franciscan created, and thus are the result of the slow dripping of silicate-rich water through cracks and crevices in caves on dry land. The mistaken moonstones at Moonstone Beach are *not* part of the plate tectonics story of California, but of a geologic process that can be found in caves around the world.

Nevertheless, almost every other pebble on Moonstone Beach is. And if one wants more evidence of the plate-against-plate collision and proof that it changed from subduction to a San Andreas–type transform fault, one has only to ask: Why is there oil in California?

Almost all of the oil in the Golden State is derived from the organics of siliceous diatom frustules. Diatoms are microscopic bivalves; frustules are the hard external cell walls covering diatoms, composed almost entirely of silica. When diatoms die, their frustules, now containing decaying organic matter, sink. Unlike radiolarians, which live far out in the ocean, diatoms thrive close to shore in shallow basins. During the middle Miocene, about 16 million years ago, there was a great proliferation of diatoms along the California coast. In part, this proliferation, which produced a thick section of organic-rich sediment, was caused by a change in the pattern of ocean currents in the Pacific Ocean—a change in pattern caused by a closing of the Indonesian Seaway as the plate carrying Australia pushed up against southern Asia. There was also a proliferation because there were the exact geologic conditions for the organic bodies of diatoms and their attendant frustules to accumulate and, later, to be buried.

Sixteen million years ago, as the transition was still under way from subduction to transform fault, a series of shallow undersea basins formed along the California coast from Newport Beach south of Los Angeles to Eel River Basin near Cape Mendocino. The geologic conditions were such that, at first, little sediment washed from the continent and over the thick accumulation of frustules; but after a few million years that changed, and those carbon-rich beds were buried. Those beds now comprise the most economically important geologic layer in California—the Monterey Formation.

The frustules in the Monterey Formation were buried at the right depth and for the right amount of time to be changed into a waxy material known as kerogen, and if the depth and time were exactly right, some of the kerogen was changed into oil, which flowed and accumulated in reservoirs.

The Monterey Formation can be found exposed at many places. One of the more spectacular is at Montaña de Oro State Park near Morro Bay, a few miles south of Moonstone Beach. Here the organic-rich rocks are a white shale. If one drives about 100 miles farther south to the seaside community of Carpinteria, one can actually see oil seeping out of the ground and onto the sandy beach—and offshore derricks where oil is being pumped out of the ground. And, of course, it is oil from the Monterey Formation that feeds the black tar to the famous La Brea Tar Pits along the 5800 block of Wilshire Boulevard in Los Angeles.

Oil production in California peaked in the mid-1970s, though today California imports two-thirds of its petroleum needs. Since the 1970s, the volume of proven oil reserves in California has dropped significantly, so that today it is less than half what it was 30 years ago.

In that regard, it should be pointed out that only a small amount of the organics in the Monterey Formation were converted into oil; the vast majority still exist as waxy kerogen. But kerogen cannot be pumped out of the ground; however, it can be extracted by a process known as fracking that requires the hydraulic fracturing of rock by pressurized liquid—an action that might induce earthquake activity—followed by the injection of chemicals to dissolve the kerogen and cause it to separate from the rock, putting it into a liquid form that could be pumped to the surface.

Where and whether or not this should be done—and what would be the economic advantages and the environmental consequences—are being hotly debated.

But this much is known: The vast reserve of oil shale that exists beneath California and the development—and continued evolution—of the San Andreas Fault will continue to have complicated effects on the region.

CHAPTER 9

TO QUAKE OR NOT TO QUAKE

Reporter: Did anyone predict last night's earthquake?
Richter: Not yet.

An attempt to use science to predict earthquakes along the San Andreas Fault began after an earthquake shook Alaska. It was Good Friday, March 27, 1964, and at 5:36 P.M. an earthquake of magnitude 9.2—the second largest ever recorded by instruments anywhere in the world*—struck southern Alaska. The rupture began under Prince William Sound and, calculations would later show, ran west for 500 miles as far as Kodiak Island. Shaking was felt over millions of square miles. In Seattle, almost 2,000 miles to the southeast, patrons sitting inside the restaurant atop the city's iconic Space Needle would later tell others that the slender tower had swayed "as in a high wind."

In Anchorage, 80 miles from the earthquake's origin, several major buildings, including the control tower at Anchorage International Airport, collapsed. The outer walls of the new JCPenney

* The largest was the magnitude-9.5 earthquake in Chile on May 22, 1960.

store fell out into the streets. In the fashionable Turnagain neighborhood, a massive landslide sent 75 houses into a nearby bay. That area has since been turned into a park—Earthquake Park. The shaking and collapse of buildings caused more than 100 fatalities in Alaska. More people were to succumb—and more damage was produced—as massive ocean waves, created by the severe shaking, swept across the Pacific Ocean.

The coastal community hardest hit was Crescent City, California. Here the local fishing fleet was swamped. Hundreds of downtown buildings were razed and more than a dozen people drowned. The level of destruction in Alaska and at Crescent City was a shock. But the question that played in the minds of seismologists in the United States immediately after the Alaska earthquake was this: What if the earthquake had happened in California?

A week later, on April 2, 1964, Frank Press, director of the Seismological Laboratory at the California Institute of Technology in Pasadena, contacted Donald Hornig, President Lyndon Johnson's chief science advisor, and suggested that a meeting be convened immediately of prominent seismologists and government officials "to discuss the problem of earthquake prediction." At the meeting, Press formed a committee to devise a prediction program based on making precise measurements that might reveal some premonitory sign—a slight warping of the Earth's surface, increased seismic activity, changes in local magnetic or gravity fields—that could warn scientists before the next major California earthquake occurred. The price of the research, Press estimated, was a staggering $137 million. He presented the program to Hornig, who promptly turned it down, citing the low annual death rate from earthquakes in the United States, which had averaged about ten deaths per year since 1906.

Undeterred, Press kept a discussion about the scientific prediction of earthquakes alive, reminding colleagues that they had a moral duty to protect society from the ravages of natural disasters. He also urged them to make earthquake prediction a legitimate scientific study and to take it away from "the purview of astrologers, misguided amateurs, publicity seekers, and religious sects with doomsday philosophies."

Ellen White, a founder of the Seventh-day Adventist Church, is credited by some as giving the first accurate prediction of an earthquake along the San Andreas Fault, though she always denied she ever made such a prediction.

She was in southern California in Loma Linda on April 16, 1906, when, as she described it, "There passed before me a most wonderful representation. During a vision of night, I stood on an eminence, from which I could see houses shaken like a reed in the wind. Buildings, great and small, were falling to the ground. Pleasure resorts, theaters, hotels, and the homes of the wealthy were shaking and shattered." Two days later, as she was on her way to a church in Los Angeles, she heard newsboys shouting: "San Francisco destroyed by an earthquake."

White had many visions throughout her adult life. The first came in 1844 at age 17 when she reported seeing a line of people making a weary and dangerous climb to Heaven, some falling "off the path into the dark and wicked world below." She also said she often saw angels. As to her 1906 vision, the day after the earthquake she told friends that more earthquakes would come and floods would follow, and that these were warnings from God that we should "not establish ourselves in the wicked cities," of which she regarded San Francisco as one.

During the next several decades, predictions of California earthquakes kept coming from other sources. In 1965, the civil defense director of Santa Barbara County actually took action after he learned that noted psychic Jeane Dixon had predicted "some tremors in July." Dixon had achieved national fame for a prediction she made in 1959 that the next person elected president would be assassinated. She later modified that prediction to say that the assassination would occur during a second term and that Richard Nixon would be the next president.

The civil defense director, Elvin Morgan, told reporters he did not want to take any chances: "If I didn't alert the people and it did happen, I would look silly." He ordered equipment removed from fire stations in case the buildings collapsed, and had police and lifeguards put on alert. Charles Richter, in Pasadena, sent letters condemning the action, saying the director had responded to "the vaporings of a crackpot." Richter also pointed out that, because of the recent earthquake in Alaska, the number of "spurious earthquake predictions" had increased. Just in the first six months of 1965, "cranks" had predicted great earthquakes for California on January 17, February 4, March 17, April 1, April 16, and all of the month of May. But, as he pointed out, so far it had been a normal year, and none of those predictions had come true.

But that was not about to discourage others, who made more predictions. By the late 1960s, the rudiments of plate tectonics and the idea that the San Andreas Fault was a major fracture running through most of California had entered the public consciousness, morphing into the idea that sooner or later the entire state would fall into the ocean. In fact, in the spring of 1969, a Calypso-style song was frequently heard on local radio stations that asked: "Where can we go, when there's no San Francisco? Guess we better get ready to tie up the boat in Idaho." It was also in the spring of 1969 that a prediction by San Francisco housewife and clairvoyant Elizabeth Steen made headlines.

Steen said she had passed her hands over a map of the United States. When her hands hovered over California, they shook uncontrollably. That caused her to leave San Francisco immediately with her husband and two children. They fled to Spokane, Washington, because when her hands had hovered over that city on the map, she "got good vibrations."

Steen's prediction prompted a steady stream of newspaper editorials and was the subject of several television documentaries, of which one emphasized that local psychics had labeled 1969 as a "doom year." Steen herself never lived to see whether her prediction came true. Before leaving San Francisco, she announced the seismic catastrophe would be in April 1969, but she died in a Spokane hospital of a blood disease on March 28 of the same year.

Years earlier, Reuben Greenspan had been described as an "earthquake prophet," a title given to him in 1935 by a *New York Times* editor. On July 9 of that year, the editor had received a letter from Greenspan stating that because the moon and Jupiter were nearly aligned in the sky, a major earthquake would occur during the next week "somewhere northeast of Australia." On July 11, an earthquake did strike in Shizuoka Prefecture, 150 miles south of Tokyo, killing nine people and injuring about one hundred. It was from that "prediction" that his fame endured.

Over the years, Greenspan, who insisted that his predictions were based on tidal calculations from solar, lunar, and planetary positions and not on prophecies, issued a series of dire warnings. Successive ones became more and more specific. And almost all foresaw the demise of San Francisco. For June 10, 1951, he predicted that an earthquake worse than the 1906 disaster would hit at 9:30 A.M. Nothing happened. His most famous prediction was his last one. In December 1972 he announced that an earthquake would strike San Francisco on January 4, 1973, at 9:20 A.M. Mayor Joseph Alioto invited Greenspan to join him for tea in the mayor's office at the prescribed time. Greenspan declined. With a week to go, the earthquake prophet recanted his prediction, saying his calculations had been wrong. He told newspaper reporters that he was leaving California and moving to Death Valley, where he would write poetry and, since the prediction of earthquakes had been solved, he would start work on the more challenging problem of the desalination of seawater.

Others continued to provide earthquake predictions. It was—and continues to be—an endeavor that many people feel qualified to do. The major requirement seems to be nothing more than a means to disseminate the message. A case in point is British science writer John Gribbin and the publication of his book *The Jupiter Effect*.

In 1974, Gribbin and co-author Stephen Plagemann argued that an "unusual" alignment of the planets in 1982 "might well trigger a California earthquake far worse than the San Francisco catastrophe of 1906." The "alignment" consisted of all nine planets, including tiny and distant Pluto, being on the same side of the sun within an arc about 90 degrees wide. Because Pluto is the slowest-moving of the nine planets, its position is the most crucial to establishing the alignment, which happens about every 180 years—which means the "alignment" is not really unusual, either astronomically or geologically. Furthermore, Gribbin and Plagemann's Jupiter effect—named because Jupiter is the most massive planet—does not explain why there was no planetary alignment during the 1906 San Francisco earthquake or during the much larger earthquakes in Chile in 1960 or Alaska in 1964 and why there isn't a worldwide increase of seismic activity every 180 years.

In 1982 Gribbin and Plagemann published a second book, *The Jupiter Effect Reconsidered*, in which they suggested that the effect had actually taken place in 1980 and had triggered the eruption of Mount St. Helens. In 1999, Gribbin admitted their argument was flawed and said, "I don't like it, and I'm sorry I ever had anything to do with it." But even today some earthquake predictors continue to use planetary positions, despite the very meager gravitational forces produced that are not even enough to affect the tides.

An ever-popular belief is that animals can somehow sense impending earthquakes and there is a wealth of anecdotal claims to support it, but no scientific evidence. Literally hundreds of species, from aardvarks to ants, have been studied. In one of the more extensive, funded for four years by the United States government, hundreds of volunteers who watched pets, farm animals, and zoo animals in California were given access to a dedicated telephone line and asked to call whenever they saw unusual behavior. Almost all of the reports filed—and there are tens of thousands—were called in *after* a felt earthquake.

One of the most persistent claims of unusual animal behavior before earthquakes has been made by James Berkland, who has tabulated the number of notices in newspapers of lost or runaway pets. His reasoning was that if dogs and cats sensed premonitory changes, those changes might make them feel unsettled and cause them to leave home. A statistical study done, independent of Berkland, of ads of missing pets in the same newspapers, which included reports of the occurrence of a magnitude-6.2 earthquake south of San Francisco, concluded there was no such correlation.

But claims that psychic energy, planetary positions, or odd animal behavior can be used to predict earthquakes persist. And, as already noted, Frank Press acknowledged this 50 years ago when he urged a serious commitment by scientists to earthquake prediction. He followed up by saying that "the forecasting of catastrophe is an ancient and respected occupation," but that it was now time for society to "part company with soothsaying and astrology" and to begin a program with "a scientifically rigorous pursuit."

Press was the son of immigrants who came from Belarus, then part of the Russian Empire. His parents left by traveling on the Trans-Siberian Railroad to the Pacific in 1916, keeping ahead of the growing turmoil as revolution swept through Russia. Somehow, they made it to China and from there to the United States, where they settled in an enclave of other recently arrived Russian Jews in Brooklyn, New York.

They found rooms in a tenement, where in 1924 their third and last child, Frank, was born. He attended public schools, which he always regarded as excellent, but where there was a problem. The smart kids were assigned to the front rows, and at first Frank was always in the back because he was unable to follow lessons. But that changed after his mother had saved enough money and bought him a pair of eyeglasses. "I couldn't see the blackboard," he would say years later. "Everything was just a big blur all the way through the first six grades, but I thought that was just the way things were." From then on, he sat in the front row—and would always be at the top of his class the whole way through.

He attended Columbia University in the 1940s, following an interest in physics and mathematics, but his attention started to shift when, while taking a geology class, he went on several short field trips to see a schist and a gneiss in Manhattan and a diabase in the Palisades.* For an advanced degree, he worked on some of the early oceanographic data being collected that would lead others to the idea of plate tectonics. For himself, he was one of the chief designers of a new type of seismometer that could record distant earthquakes. In 1957, because of his new type of seismometer, he was appointed director of the Seismological Laboratory in Pasadena; then eight years later, because of his managerial skills, he was appointed head of the Department of Earth and Planetary Sciences at the Massachusetts Institute of Technology in Boston. Press was a rising star—and he would rise much higher.

During his career, he worked on a wide range of problems in seismology, both practical and theoretical. He was involved in determining how to verify that the Soviet Union was conforming to the conditions of the Test Ban Treaty signed in 1963 to limit the testing of nuclear weapons. He was among the first to study and explain how the Earth "rang like a bell" when a sufficiently large

---

* These rocks are easy to find in the New York area. The Manhattan Schist can be seen exposed in J. Hood Wright Park, and the Fordham Gneiss (one of the oldest rock formations in the world, dated to have formed 1.1 billion years ago) is exposed at Inwood Hill Park, both in northern Manhattan. The choice of where to build skyscrapers is limited by where these rocks are close to the surface. The Palisades Sill is a diabase that intruded about 200 million years ago and that forms the cliffs along Henry Hudson Drive in New Jersey directly west of Manhattan Island, best seen just north of the interstate approach to the George Washington Bridge.

earthquake—like the 1960 Chile earthquake or the 1964 Alaska earthquake—caused the entire planet to vibrate. But the idea of earthquake prediction held a special fascination, one that was not deterred by the inability to get a national program of seismic research started after the great seismic catastrophe happened in Alaska in 1964. It was a matter of waiting for the right opportunity. And the next opportunity came early on the morning of February 9, 1971.

It is known as the earthquake that woke Los Angeles—both literally and figuratively. At 55 seconds before 6 A.M., the ground began to vibrate. The strong shaking lasted 12 seconds. Then at 6:01 A.M., there was a second shock, smaller than the first; then minutes later four more shocks so close together that they felt like one long earthquake. In all, the series lasted 5 minutes and 11 seconds. In that time, the ground surface ruptured and two major hospitals suffered extensive damage in the Sylmar District of the San Fernando Valley in California.

Some of the surface rupture can still be seen today, though it has been smoothed over by bulldozing, and the mile-long scarp that was formed is often difficult to find, now covered by houses or small commercial buildings. The best place to see it is on Glenoaks Avenue south of Hubbard Street. Here the drive-through lane of a fast-food restaurant runs right along the base of a three-foot-high step dividing the restaurant from its parking lot. That step is where the ground ruptured and rose in 1971.

Of immediate concern after the earthquake were the two hospitals. Forty-nine people died during the collapse of the Veterans Administration Hospital, built in 1925 and not designed to withstand earthquakes. The Olive View Medical Center, which was designed to withstand earthquake shaking, had opened a month earlier. Fortunately, it was unoccupied when the four wings of the five-story structure pulled away from the central structure and three stair towers toppled.

Elsewhere, the shaking caused 12 overpass bridges to crash onto freeway lanes. Fortunately, the earthquake struck in the early morning hours and few vehicles were on the freeway, though Clarence Dean of the Los Angeles Police Department died when he was unable to stop his motorcycle before it ran off a partially collapsed bridge. The greatest concern—even outdoing the damage to the hospitals—was the failure of the Lower Van Norman Dam.

Built in 1916 to hold water coming into the Los Angeles area via the Owens Valley aqueduct, the Lower Van Norman Dam was an earth-fill dam consisting of a clay core covered with a thick layer of sand. The shaking had caused the upstream part of the sand layer and part of the clay core to slide into the water reservoir, leaving only four feet of freeboard between the water level and the top of the now-damaged dam. Eighty thousand people who lived downstream were evacuated immediately. The dam was repaired and the people returned.

The 1971 San Fernando earthquake, which at magnitude 6.5 is regarded as a moderate event, was the first damaging earthquake in the Los Angeles area since 1933. It showed how inadequately prepared California was for a truly large earthquake. It also set off a renewed interest in reducing hazards related to earthquakes, and brought forth the harsh reality that while the annual average of deaths by earthquakes might statistically only be ten per year, it would only take one catastrophic incident to make that number rise exponentially.

On November 21, 1972, the Hospital Safety Act was signed, which required the same strict building standards be applied to hospitals that were already—as a result of the 1933 Long Beach earthquake—applied to public schools. A month later, another legislative act was signed that prohibited the construction of new houses within 500 feet of an active fault—to lessen the possibility that an earthquake rupture might form in someone's backyard or through their bedroom. And in December 1974 the California Seismic Safety Commission was established to advise the governor on future programs needed to reduce earthquake risk. But there was nothing about prediction.

In 1971, the failed forecast by psychic Elizabeth Steen was still fresh in people's minds. And for years after 1971, Reuben Greenspan kept telling anyone who would listen that he had predicted the earthquake two days before it occurred. By the time the Seismic Safety Commission was organized, *The Jupiter Effect* was selling well, its contents regarded by some as a justifiable end to a hedonist society.

And yet, though Frank Press tried to find support, the 1971 San Fernando earthquake was still not enough to direct interest toward earthquake prediction. Instead, as Press now realized, such a redirection would require a truly colossal event and the startling claim that some credible person—or persons—had predicted it. That claim came in February 1975 when rumors circulated that Chinese scientists had successfully predicted a destructive earthquake 24 hours before it occurred, and as a result forced the evacuation of a large city, thereby saving hundreds of thousands of lives.

Haicheng is located about 300 miles northeast of Beijing in Liaoning Province. In 1975 it had a population of about 1,000,000. It is a mining center, touting itself today as "the talc and magnesium capital of the world." To seismologists, it is the site of the only successful prediction of a major earthquake—with qualifications.

Beginning in September 1974, many small local earthquakes were recorded by Shipengyu Earthquake Observatory, located 15 miles from Haicheng. On December 22, the first widely felt event, a magnitude-5.2 earthquake, was recorded by the observatory. Six days later, as the seismic activity continued, local government officials had their first meeting to discuss the possibility of a major earthquake. Over the next month, they issued four specific predictions. All were false alarms.

At 6:00 P.M. on February 3, seismic activity increased dramatically. By the next morning, when at least one earthquake was rocking the ground every few minutes, the head of the local earthquake office, Cao Xianqing, called a meeting of government officials. He stood before them and said: "A large earthquake may occur today during the day or the night." He asked that a formal announcement be made to warn the public.

That afternoon, February 4, the government ordered all businesses and factories to close. It also suspended all public meetings and sports activities. A movie operator who, after talking to workers of the Shipengyu Observatory, was convinced that a major earthquake was imminent, decided to show movies outdoors all night to attract people away from their houses. Other people, concerned about the continued ground shaking, left their homes and prepared to sleep outside, even though it was midwinter.

At 7:36 P.M. a magnitude-7.0 event, comparable to the San Fernando earthquake, shook the ground. Most of the buildings of Haicheng and in the surrounding communities collapsed. More than 2,000 bridges were damaged. Hundreds of breaks occurred along gas pipelines. Yet, considering the severity of the event, the number of fatalities was extremely low. Some investigators, after viewing the destruction, guessed that the death toll could have been as high as 150,000, but the actual number was barely 2,000. By comparison, the same night, 6,578 people were treated for frostbite and 372 died from freezing.

So the question is: Did Chinese officials actually *predict* the earthquake that led to the evacuations? The answer is: maybe.

It has never been clear what the workers at Shipengyu Observatory told the movie operator that made him show movies outdoors, hoping to keep people out of their houses. Nor is it known whether those same observatory workers were ever told that Cao Xianqing thought a major earthquake was possible.

Xianqing has been interviewed several times and he confirms that, hours before the event, he did say repeatedly to a few colleagues at the government earthquake office—not Shipengyu Observatory—that a major earthquake would occur before 8:00

P.M. on the night of February 4. But why did he make such a prediction? When pressed, he answered that he knew of a book, printed in the 18th century, that said, "Excessive autumn rain will surely be followed by a winter earthquake." The previous autumn had been exceptionally rainy, and according to the Chinese calendar, February 4 was the last day of winter. An earthquake swarm did begin on the evening of February 3, and when he made his first prediction the next morning, earthquakes were being felt every few minutes.

Also, some large towns were evacuated, such as Dashiqiao, where, out of a population of about 72,000, only 21 people died, though two-thirds of the buildings collapsed. Who ordered the evacuation of Dashiqiao remains unclear.

After the earthquake, several possible precursors were identified. The land around Haicheng had risen slowly as much as two inches since the summer of 1973. There were changes in the level of local water wells and in the color of water in those wells. The rate of emission of radon gas increased. And there was anomalous animal behavior. Of the last, the most noted was the appearance of snakes that should have been in hibernation; however, snakes first appeared about 24 hours before the main earthquake, as soon as the lesser shaking began on February 3, so seismic activity had already started before the snakes were seen.

Nevertheless, the prediction of the 1975 Haicheng earthquake was—and still is—touted as successful. It caught the attention of Frank Press and others, in part because for more than a decade scientists in the Soviet Union had been claiming that they, too, knew how to predict earthquakes.

The strategy followed by Soviet scientists was completely different from that taken by scientists in China. In the Soviet Union, earthquake prediction was based on precise measurements. In particular, Soviet scientists had noticed that the speed of seismic waves slowed down when passing through a region where a large earthquake later occurred. And there was laboratory evidence and a theory to support it.

When a rock sample is squeezed in a vise in a laboratory, tiny cracks form close to the line where a fracture eventually develops. If one taps the sample with a hammer, the speed of the disturbance produced by the hammer traveling through the sample is slowed by the tiny cracks. The same thing happens, so it was argued, in nature. When two tectonic plates slide against each other, tiny cracks form in the Earth's crust and, eventually, a large fracture—an earthquake—occurs. The presence of those cracks explains the slowing of seismic waves before an earthquake observed by Soviet scientists.

Furthermore, the formation of tiny cracks in a laboratory sample causes the sample to expand, or dilate, slightly. That could explain why the ground surface rose two inches a year or so before the 1975 Haicheng earthquake—and why, as reported by Japanese scientists, sections of the coastline have risen a few inches before at least some major earthquakes, such as the magnitude-7.5 earthquake near Niigata in 1964.

It all seemed to fit together. Laboratory tests could explain why Chinese and Soviet scientists could predict earthquakes, and there was a theory—dilatancy theory—that supported the claims. By the mid-1970s, the routine prediction of earthquakes seemed to be possible in the near future—perhaps, as some scientists including Frank Press said, as soon as five years.

Los Angeles was now the third largest city in the nation. It was only a matter of time, Frank Press and others realized, until some truly colossal event—similar to the 1857 earthquake—was unleashed along the San Andreas, an event that would be many times more powerful and more destructive than the 1971 San Fernando earthquake. Given the reported successes in China and the Soviet Union, it should be possible to recognize a sign that such a catastrophic event was imminent. Surprisingly, such a sign came sooner rather than later.

In the aftermath of the 1971 earthquake, scientists combed through measurements of land surveys to see if there was any evidence that the ground surface had moved in some systematic way before the earthquake. No evidence was found. But something more intriguing was uncovered.

Since 1960, it seemed that a broad region, more than 4,000 square miles in extent, had risen, centered on the San Andreas Fault. The maximum uplift was ten inches near Palmdale, so this broad uplift soon became known as the Palmdale bulge.

In April 1976, at a scientific meeting held in Washington, D.C., Press gave an address entitled "A Tale of Two Cities." In it, he reminded his audience that a year earlier Chinese scientists had accurately predicted an earthquake, in part because they had recognized a broad uplift of the land around Haicheng city. Now a similar bulge had been found along a stretch of the San Andreas Fault north of Los Angeles. This "worrisome uplift," as he called it, might or might not be a sign of an impending disaster, but "precautionary measures will not be in vain." And he added, "The San Andreas Fault will undoubtedly rupture again."

Later at the same meeting, a young researcher from the Seismological Laboratory in Pasadena, James Whitcomb, gave a paper in which he reported that he had uncovered a slowing of seismic waves that passed through the San Andreas Fault north of Los Angeles. Three years earlier, in November 1973, Whitcomb had found a similar slowing of seismic waves east of Los Angeles near Riverside and from that had predicted a magnitude-5.5 or greater event within three months. A magnitude-4.1 earthquake did occur in the designated area on January 30, 1974. Now, in April 1976, on the basis of the experience in the Soviet Union, he expected a magnitude-5.5 to -6.5 earthquake north of Los Angeles during the next 12 months.

The *Los Angeles Times* interviewed Whitcomb repeatedly, explained the dilatancy theory, and gave advice on how people should prepare themselves for the coming earthquake. A city councilman in Los Angeles threatened to sue Whitcomb for hurting property values in the city. The *San Francisco Chronicle* joined the *Times* in publishing editorials supporting a federally funded prediction program.

But by summer, the optimism started to wane.

On July 28, 1976, a magnitude-7.6 earthquake struck the city of Tangshan, 200 miles southeast of Haicheng, killing 650,000 people. It was the greatest earthquake disaster of the 20th century. No broad region of uplift or slowing of seismic waves was noticed before the earthquake. It brought into question whether the Haicheng example had been a fluke. Clearly, the prediction program in China was not reliable enough to prevent future disasters.

Thus there was more pulling back of earthquake prediction.

Whitcomb continued to analyze the speeds of seismic waves that passed through southern California, coming to the conclusion that they did not follow the pattern required by the dilatancy theory. In December 1976, he publicly withdrew the prediction.

Meanwhile, additional surveys across the region of the Palmdale bulge showed curious movements. The bulge was rising in places, while dropping in others, without any relation to seismic activity. By 1979, most of the bulge had disappeared, giving rise to questions about whether the bulge ever existed, and whether the original claim of a broad uplift might have been due to subtle measurement errors.

But back in 1976, there was real concern that the bulge was a harbinger of a major earthquake that would devastate Los Angeles. On January 14, 1976, Press gave a presentation at the White House to administration officials. As if to emphasize what could happen to Los Angeles, three weeks later an earthquake rocked Guatemala, killing 23,000. Now that something had happened in America's own hemisphere, political wheels were set in motion toward starting a national program of earthquake research. The final assurance that the program would be realized came in March 1977 when the new president, Jimmy Carter, named Frank Press to be his science advisor.

The Earthquake Hazards Reduction Act was passed unanimously by the Senate and by a small majority in the House and became law in October 1977. Overnight, the amount of money available for earthquake research in the United States tripled. The main purpose of the act, however, was not to predict earthquakes but to reduce the impact of future ones. In that regard the program was a success, advancing engineering studies of the effects of strong seismic shaking on buildings and bridges, identifying previously unrecognized fault strands, examining the history of the San Andreas Fault and many related topics. But the prediction of earthquakes was definitely in the background until 1982, when the program was up for renewal and members of Congress wondered what had happened to earthquake prediction.

Senator Harrison Schmitt, a former astronaut and the only scientist to land on the moon, conducted hearings to determine if sufficient progress had been made under the Earthquake Hazards Reduction Act to predict earthquakes. After hearing from several experts, including Frank Press, the Senator decided there had not. He threatened to terminate the entire program unless an "earthquake prediction system" was in place somewhere in the United States "within four or five years."

There was a rush to decide where such a system might be located. Where in the United States was a major earthquake most likely to occur within the next few years? A consensus grew: The most likely place was along a middle strand of the San Andreas Fault at Parkfield, California.

The San Andreas Fault can be divided into three main segments. The northern segment runs from Cape Mendocino to San Juan Bautista—the part of the fault that ruptured in 1906. The southern segment begins around Cholame, just north of Carrizo Plain, and runs south, eventually forming the southern boundary of the Mojave Desert, continues through Cajon Pass and San Gregorio Pass, can be picked up 20 miles east of Palm Springs in Coachella Valley, and ends at Bombay Beach on the east side of the Salton Sea. The northern half of the southern segment—from Cholame to Cajon Pass—ruptured in 1857; the southern half of the southern segment—from San Gregorio Pass to Bombay Beach—did so in about 1690. So all of the San Andreas Fault has broken during a major earthquake in the last few hundred years except for a short middle segment that runs from San Juan Bautista to Cholame and includes the ranching community of Parkfield. This 150-mile segment of the San Andreas Fault is distinctly different from the other parts of the fault: Here the fault is slowly and continuously sliding.

Ten miles south of San Juan Bautista is DeRose Vineyards. It is a family-owned business where the winemaking and tasting room is located in a large building with a concrete floor and metal walls and roof. On the day I visited, I identified myself as an earthquake tourist. The person who was pouring the wine pointed immediately to the center of the building and said, "It's over there."

Here the trace of the San Andreas Fault is all too apparent. Running along the floor is a line of broken concrete slabs, up to a foot across, that extends the full length of the building. Where the fault runs beneath a metal wall, the wall has been sheared apart, the two halves now standing as much as two feet apart. Broken ends of twisted rebar are exposed where the metal wall once connected to the concrete floor. A plaque attached to a wall in the center of the DeRose Winery building proclaims the San Andreas Fault at this spot to be a registered natural landmark.

If one drives south of DeRose Vineyards, one can find sets of cracks running diagonally across the pavement. These, too, are the San Andreas Fault. They are visible, as is the slow destruction of the winery at DeRose Vineyards, because along this segment the fault is always sliding. And the sliding can be found as far south as Parkfield, where the fault runs under a bridge. As one might expect, the bridge has a distinct bend over the exact place where it crosses the San Andreas Fault.

The slow sliding is known as seismic creep, caused in part by a constant jitter of small earthquakes. At DeRose Vineyards, the fault slides about an inch a year. At Parkfield, it is half that amount, which means occasionally the Parkfield section has to catch up. It does so with a jolt—a moderate earthquake.

Six times—in 1857, 1881, 1901, 1922, 1934, and 1966—the Parkfield section has surged forward. Each event has been nearly identical in size—corresponding to a magnitude-6 earthquake—and each successive event has occurred, on average, 22 years after the previous one. Moreover, there seemed to be definite precursory signs before the last two events. The main shocks in 1934 and again in 1966 were preceded 17 minutes by a strong foreshock that was felt over a wide area. Furthermore, an irrigation pipe that crossed the rupture zone separated nine hours before the 1966 event. All this gave credence to the idea that the next Parkfield earthquake might be predicted.

In 1985, a panel of 12 scientists, formally known as the National Earthquake Prediction Evaluation Council, endorsed a Parkfield prediction, saying that there was a 95% chance that a magnitude-6 earthquake would occur along the Parkfield section of the San Andreas Fault by 1993. A dense network of instruments was installed in the hopes of trapping the earthquake, to detect precursory signs that might occur in seismic patterns, in ground movements, in electric or magnetic fields, in radon-gas emission, or in the chemistry or level of water wells. And then people waited.

Twice, an "A"-level alert was issued, on October 19, 1992, and again on November 14, 1993. Both alerts were triggered after felt earthquakes, similar in size to what preceded the 1934 and 1966 events, occurred. Both times there was an increased awareness that the predicted earthquake might occur within the next 72 hours. California state agencies and emergency services were notified. And both times . . . nothing happened.

The year 1993 came and went, and no earthquake. Then 1994, 1995, and so on. Finally, at 10:15 in the morning on September 28, 2004, a magnitude-6 earthquake ruptured the Parkfield section of the San Andreas Fault. The predicted earthquake had occurred. Or had it?

There were important differences between the events of 1934 and 1966 and the one that occurred in 2004. First, in 1934 and 1966, the ground rupture began north of Parkfield and propagated south. In 2004, it was in the opposite direction: The rupture started south of Parkfield and propagated north. More important, in the dense network of instruments there were no precursors recorded minutes, hours, or days before the event. There was no foreshock or increase of seismic activity before the event. There was no damage to irrigation pipes. There was no measured change in electric or magnetic fields or in chemistry or level of water wells. Most disconcerting, there was no measured ground movement: There was no warping or rise or fall of the ground surface. There was no underground compression or slight expansion of rock—no dilatancy—and this could be measured with great precision.

Five instruments known as borehole strainmeters were installed within a few miles of where the 2004 rupture formed. Essentially, each instrument consists of a fluid-filled bag stuffed deep down a borehole. If the surrounding rock is compressed or stretched by a tiny amount—equivalent to taking a 100-mile-long rigid bar and compressing or stretching one end by the diameter of a human hair—the bag undergoes a small compression or expansion. But no change was recorded for weeks to seconds before the earthquake. As far as anyone can tell, the 2004 Parkfield earthquake was a spontaneous event.

The Parkfield experiment was successful in identifying *where* an earthquake would occur and *how* big it would be, though the all-important *when* was missed; the event came 12 years too late. Which raises the question: Will it ever be possible to predict earthquakes?

The answer, as it is seen today, is: maybe.

The quickest way to start an argument in a room filled with seismologists is to bring up the question of earthquake prediction. Passions can escalate from heated to downright nasty.

An extreme example occurred during the prolonged debate about the Palmdale bulge and whether an earthquake was imminent along that section of the San Andreas Fault. Ross Stein at the United States Geological Survey had argued for years that the bulge was real, but then, after more data was available, he changed his mind. His conversion did not sit well with at least one colleague because *twice* Stein found a bag of dog excrement in his office mailbox.

Fortunately, such juvenile behavior is rare and should not distract from the serious business of trying to predict earthquakes.

Furthermore, the question of earthquake prediction can be reduced to a more tractable and straightforward question: What triggers a large earthquake?

Imagine this: Initially, an earthquake fault, such as the San Andreas, is relatively quiet. Only a few small earthquakes are occurring, popping off, in a familiar analogy, like kernels of heated popcorn. Then the popping of one earthquake kernel happens to set off more kernels, and those set off more kernels until there is an explosion, or cascade, of kernels popping, a rupture forms, and a large earthquake is produced.

Or imagine this: The lower region of the San Andreas Fault is slowly sliding—without earthquakes—because here the rocks are hot and plastic and are driven to slide smoothly by the slow and ever-constant movement of the Pacific and North American plates. As the slipping region grows, the sliding accelerates until it reaches a critical speed to where a rupture in the brittle overlying rock forms, and a large earthquake is produced.

In the former case—the cascade model of earthquake kernels—the beginning of any large earthquake is no different from the beginnings of countless small ones, which means it is impossible to ever predict large earthquakes.

In the latter case—the pre-slip model—a long process occurs that prepares the San Andreas for a sudden and major slip. In that case, earthquakes might be predicted if we can figure out how to measure the slow sliding and subsequent buildup.

Which idea is true—or whether earthquakes work in some other manner entirely—is still the focus of much research today and is hotly debated. But this much is true: When there is a major earthquake, the probability of another major earthquake happening soon after in the same region goes way up. Once the Earth's crust starts to adjust to the slow buildup of pressure between the tectonic plates, that pressure may not be relieved simply as a single large event but rather as one—or more—major earthquakes happening in a short time period.

To put this in concrete numbers: History shows that whenever there is a major earthquake in California, say a magnitude-6 event—which can do substantial damage—there is a 1 in 10 chance that another earthquake of equal or greater magnitude will happen in the same general area *within the next three days.*

This leads to a practical concern. After a major earthquake, people should brace themselves for an equal or larger event. Emergency services, such as fire and police stations and hospitals, need to prepare for additional injuries and for the disruption of still more roads and utilities. And those who are attempting to rescue people who are already trapped under debris should be aware that a larger earthquake could strike and a greater catastrophe could happen.

CHAPTER 10

ANCIENT TREMORS

We can't, of course, know exactly what the San Andreas Fault has in store for us.

—Kerry Sieh, on future earthquakes, 1981

One of the enduring mysteries of southern California is the annual return of swallows to the old Spanish mission at San Juan Capistrano. The return is part of a 15,000-mile migration the cliff swallow makes from the west coast of South America to the west coast of North America and back again. The birds, of course, congregate and nest at many sites in North America, though San Juan Capistrano is the most famous, memorialized in song and in a movie and illustrated and told in countless tourist brochures. But one wonders: Why San Juan Capistrano?

The cliff swallow, *Petrochelidon pyrrhonota*, needs three things to thrive. It needs an ample source of insects to eat and a source of mud to build a nest. Both can be found in many places in California. But the third requirement is a challenge. The cliff swallow prefers to build nests under cliff overhangs or at cave entrances. Neither can be found near San Juan Capistrano, but there is something just as good. There are high walls and arched insets of a great stone church—a church that was ruined by an earthquake.

Work was completed on the great church at San Juan Capistrano on September 8, 1806. At the time, it was the only church in Spanish California not constructed out of adobe. Instead, blocks of cut sandstone were used for the walls and vaulted ceilings and for a high bell tower that stood next to the main entrance. It was said the tower could be seen from distances as great as 10 miles, and that the sound of its bells could be heard from even farther. This marvelous structure, which took nine years to build, collapsed in less than a minute on the morning of December 8, 1812.

On that date, the ground shook twice. The first shock caused the bell tower to sway. Parishioners who were inside the church, mostly Native American women and their children who were there for the first Mass of the day, felt the shaking and heard the bell ring. They ran for the main entrance, but the shaking had jammed the door. A second shock, stronger than the first, amplified the swaying of the bell tower and caused it to fall onto the church through the stone roof. Forty people were killed. Eventually the debris was cleared away and the bodies recovered and given burials, but the walls and the arches that still stood were thought to be too precarious to tear down. So the ruined walls and arches were left standing. And the swallows took up annual residence.

The same earthquake also damaged the church missions at San Gabriel, San Fernando, and San Buenaventura, the last one 100 miles west of San Juan Capistrano, which indicates this was felt over a large area and therefore was a major event. But where was the earthquake located? What was its epicenter?

Such a complex lattice of faults runs through southern California that it might seem impossible to ever know exactly which fault had slipped on December 8, 1812, and damaged and caused fatalities at the great stone church at San Juan Capistrano. But in 1975, an examination of tree rings revealed the source.

The fact that large earthquakes can affect the growth of trees is well documented. Severe ground shaking can bend a tree far over, breaking many limbs. Ground rupture can sever roots, retarding growth. In 1906, near the northern end of the San Andreas Fault near Fort Ross, several large trees were actually split apart by the ground rupture, but survived and lived to show the ordeal to future scientists.

In 1975, cores were drilled into 65 Jeffrey pines standing close to the San Andreas Fault northwest of Cajon Pass. The purpose was to see if the rings of any of the trees indicated retarded growth after the 1857 earthquake. The cores of nine of the trees did show a narrowing of tree rings after 1857. And there was more.

There was another narrowing of tree rings that began between the end of the growing season in the fall of 1812 and before renewed growth in the spring of 1814. Could this be evidence that the San Andreas Fault, where it passes through Cajon Pass, had been the source of the December 8, 1812, earthquake? If so, there should be additional evidence of the earthquake elsewhere along the fault. But how might it be possible to cut into and examine a fault and, in all the complexity, see the effects of a single earthquake?

Imagine you are a baker. A customer comes into your store and orders a six-layer wedding cake. You are known to be methodical and to always begin baking at noon. You are also poor and have only one cake pan.

You prepare the batter for the first layer, bake it, and pull it out of the oven at precisely 1:00 P.M. and set it on a plate. You prepare the next layer, pull it out at 2:00 P.M., set it atop the first layer, then prepare the third layer, pull it out at 3:00 P.M., and so forth.

Sometime while the fifth layer is baking an earthquake happens and the unfinished four-layer cake falls to the floor. Fortunately it does not crumble, but it splits into two parts. You scoop it up and set it back on the plate, adding the fifth layer at 5:00 P.M. and the sixth layer an hour later. You then cover the whole cake with frosting and deliver it to the wedding, hoping that no one will be able to figure out what happened.

You are unaware that the bride is a geologist.

When she cuts the cake, she notices a crack that extends through all the lower four layers but not into the upper two. She also knows of your methodical baking habits and easily figures out that whatever calamity happened to the lower four layers to split the cake must have occurred between 4:00 P.M. and 5:00 P.M. That, in essence, is the foundation of paleoseismology.

Paleoseismology is the study of the geologic effects of past earthquakes. Such effects include the fracturing, warping, folding, or sliding of sedimentary layers that were laid down at the bottom of swamps, rivers, or lakes. The key is finding a place where the layers were deposited continuously for many, many years and where they lie over an active fault. In principle, the work of deciphering these layers is straightforward—as the cake analogy demonstrates—but in practice, it is tedious and can best be described as trying to produce a highly detailed geologic map of an area that can be covered with a bed sheet. It requires distinguishing the limits of myriad mud, sand, and gravel deposits, then tracing hairline cracks through them. It also requires that layers of organic material, usually peat, be present so that radiometric dating of the carbon can be done to determine when the various hairline cracks formed. It is meticulous, highly frustrating work. At times, it seems impossible—and that's because it almost is.

Two early attempts at paleoseismology were tried immediately after two moderate earthquakes in southern California, after the 1968 Borrego Mountain earthquake west of the Salton Sea and after the 1971 San Fernando earthquake. Ruptures of both tremors broke the surface, and in both cases bulldozers were used to dig trenches across each break. The results were disappointing.

The problem, as one investigator put it, was that in the exposed walls of the trenches, it was "extremely difficult to see" where the fault breaks were because they passed through "massive unbedded material ranging from silt to clay to coarse bouldery sand and gravel." What was needed, recommended the same investigator, was someone who was willing to devote years to digging, scraping, and carefully brushing the walls of such trenches to reveal the barely discernible cracks and offsets of mud and sand layers that were caused by recent earthquakes.

Enter Kerry Sieh.

Sieh's fascination with the San Andreas Fault began when he was a student at Stanford University in the 1970s. He decided he wanted to know more about the last great earthquake to occur along the San Andreas Fault before 1906: the 1857 Fort Tejon earthquake. Surprisingly little was known about the earthquake when Sieh started his work, the shaking having occurred when the population of the greater Los Angeles area was just a few thousand people.

Sieh tracked down more than 60 personal accounts of the earthquake, mostly found as brief newspaper reports, as well as several personal letters and a few unpublished memoirs that described the event. Then he set off and within a year had hiked or bicycled most of the 200 miles of the rupture, looking for indicators of where the fault had slid, focusing much of his attention on where stream channels had been offset. Next, he decided he needed to dig to determine exactly where the fault trace was located and exactly how much the fault had moved in 1857. At that notion, his Stanford professors balked.

It was doubtful, so they said, that the rupture of an earthquake, even one as large as in 1857, would leave a readable record after more than 100 years. Besides, they said somewhat quietly, Sieh was proposing to spend years examining just the thin surface layer that most geologists ignore, instead of delving deeper into bedrock, where the clues to the fundamental problems of the growth of mountains, the origin of continents, and the history of ocean basins would be found.

But Sieh, with the stubbornness of a brilliant student, was undeterred. He started his project armed at first with only a shovel. But where should he dig? Here his yearlong reconnaissance of the San Andreas Fault paid off. He decided on Pallett Creek, a few miles west of Valyermo.

Pallett Creek is an ephemeral stream that flows down the north side of the San Gabriel Mountains and onto the Mojave Desert. It is at the base of those mountains that the creek crosses the San Andreas Fault, a setting favorable for the repeated deposition of mud and gravel and the development of the occasional layer of peat. Taking advantage of a 30-foot-deep erosional trench made by the creek—and, eventually, with his work augmented by use of a backhoe and a bulldozer—Sieh dug into the ground and exposed a profile of the very upper reaches of the San Andreas Fault.

About two feet below the ground surface, he uncovered a former surface layer that had a broad warp in it. By carbon age dating of decomposed plants that once grew on the former surface, Sieh was able to determine that the warping had been caused by the 1857 earthquake. A foot deeper there was another disruption, this one caused by the 1812 earthquake—confirming that the earthquake that damaged the church at San Juan Capistrano had occurred along the San Andreas Fault. And greater surprises lay beneath.

By tracing disrupted layers, Sieh identified the occurrence of seven more earthquakes, determining the timing of these events also by carbon dating. The next older event before 1812 had occurred in about 1500. The earliest one he exposed had occurred in about 730 A.D. It was a spectacular discovery because the record of individual earthquakes for the San Andreas Fault had been pushed back from the 200 years of historical records to a paleoseismic record that covered nearly 1,500 years.

But the historic and prehistoric earthquakes Sieh identified at Pallett Creek represented only one point along the hundreds of miles of the San Andreas Fault. Certainly, there must be similar records at other places. And indeed there are.

In the decades since Sieh first pushed a shovel into the soil at Pallett Creek, 22 sites have been uncovered where paleoseismic work has been done. They extend from the northernmost site at Shelter Cove just south of Cape Mendocino to Salt Creek on the east edge of the Salton Sea. But the record at Pallett Creek is one of the best because it has the second longest record, exceeded only by a slightly longer one at nearby Wrightwood, located halfway between Pallett Creek and Cajon Pass. Here the record of earthquakes along the San Andreas Fault goes back an additional 200 years, the earliest event occurring in about 530 A.D.

Not surprisingly, when the records of different sites are compared, the effect of the same earthquake is seen at more than one site. In the last 1,500 years, there have been at least three earthquakes that broke the same section of the San Andreas as happened in 1857. These three earlier events occurred in about 1360, 965, and 655 A.D. We also know that there was a flurry of four events that broke through Wrightwood near Cajon Pass between 600 and 800 A.D. In fact, when looking at records from all 22 sites, it seems that after three or four centuries of quiescence, seismic activity resumes and a series of major earthquakes occurs along the entire length of the fault. It took a record that extended back more than 1,000 years into prehistory to give us a long enough sequence of earthquakes to make these correlations; relying on a history of barely 200 years was not taking into account enough time to understand how activity shifts along the San Andreas Fault.

In short, thanks to Sieh's findings, it was revealed that major earthquakes along the San Andreas Fault are not random but occur as clusters—an important point that will be returned to later in this book.

The success of paleoseismology in California, pioneered by Sieh, set off a revolution in seismological research. In Italy, for example, more than 100 trenches have been dug across dozens of active faults, revealing evidence for scores of previously unknown individual earthquakes dating back as far as 6,000 years. So far, the longest and most complete paleoseismic record comes from a region of the Dead Sea in the Middle East where thinly bedded sediments laid down by a former, much larger sea contain evidence of individual earthquakes going back 50,000 years.

A study of those sediments confirms that a major earthquake happened in the Dead Sea region in 31 B.C., an earthquake that, according to historical documents, damaged the Second Temple in Jerusalem and Herod's Winter Palace in Jericho. Archaeological evidence suggests it also damaged the city of Qumran on the West Bank, disrupting the city's water supply and causing people to leave. It has also been suggested that the disruption recorded in the bedded sediments of the Dead Sea in 31 B.C. might have been associated with shaking so strong that it closed off a cave near Qumran, sealing for nearly two millennia an archaeological treasure known as the Dead Sea Scrolls.

Paleoseismology has solved a host of other problems. For example, in Greece there has been a long-running controversy about the sanctuary of Delphi, where in ancient times a series of specially trained women sat inside a temple and answered questions with prophecies. According to Plutarch, who ran the sanctuary for many years, a priestess uttered an oracle after inhaling vapors that rose from a chasm in the earth beneath the spot where she sat. For years, the existence of the chasm was dismissed because archaeologists could not find any trace of it. But paleoseismologists, who examined the site with differently trained eyes, have identified an active fault that passes directly under the temple. Groundwater is easily accessible near the fault because the fault gouge is impermeable. Furthermore, they have also found a large body of limestone with bitumin, a petrochemical that, when dissolved by groundwater, might have released ethylene. Inhaling ethylene produces euphoria—which is why it was once used as an anesthetic—and that could explain the trance-like state of the Delphic oracles.

Another mystery that paleoseismology has shed light on is the apparition of Archangel Michael at Monte Sant'Angelo on the Adriatic coast of Italy in 493 A.D. According to the one person who claimed to have seen the archangel, the first Bishop of Siponto, Michael's appearance coincided with *immenso termore,* an immense earthquake. Paleoseismology has confirmed that a large earthquake did occur near Monte Sant'Angelo in about that year; in fact, it was the largest earthquake to strike that part of Italy in the last few thousand years. The tools of paleoseismology can also explain the material evidence left of Michael's apparition as described by the Bishop of Siponto: the appearance of a giant footprint at the entrance of a sacred cave (where, later, the Footprints Altar was built). Trenching shows that the "giant footprint" was a local subsidence of the land related to the surface rupture that caused the earthquake.

On September 9, 1349, a strong earthquake shook much of the Italian peninsula—an account of which was written by the poet Petrarca—causing many major buildings to collapse, including the outer south wall of one of the iconic monuments of the ancient world, the Colosseum in Rome, giving the monument its famous asymmetric look. But where did that earthquake originate?

Paleoseismological sleuthing, which involves trenching as well as—in this case—the examination of aerial photographs and the reading of numerous Annals and Chronicles from the Middle Ages, reveals that a 15-mile-long rupture formed at the base of Matese Massif in the southern Apennine Mountains about 50 miles southeast of Rome in 1349. But why did the south wall of the Colosseum collapse, and not the north wall?

Trenching into the base of the Colosseum has shown that the northern half of the monument lies on firm ground, while the southern half was built on loose sediments that filled a former tributary of the Tiber River. The ground shaking in 1349 was amplified by the sediments, which, because they were not compacted, behaved like quivering jelly, causing the outer south wall of the Colosseum, having no firm base, to collapse. It is also recorded that the failure of the south wall set off a quarrel between Pope Clement VI and some powerful Roman citizens over who had the right to collect the fallen stones and use them in construction projects.

The use of paleoseismology to solve archaeological problems and theological quandaries shows that the techniques of the new science are not limited to just an examination of individual geologic layers and how they may be disrupted. It is much broader than that. A case in point is how a 3,000-year history of major earthquakes was determined for the offshore region of northern California and the Pacific Northwest—and how those earthquakes relate to the San Andreas Fault.

This is the region of the Cascadia subduction zone where two remnants of the Farallon plate—the Juan de Fuca and the Gorda plates—are sliding under the North American plate. The sliding does produce large earthquakes, the latest on January 26, 1700. That earthquake and its predecessors certainly disrupted geologic layers, but these layers lie under the sea and are not easy to study. Thankfully, there is another record of these earthquakes.

Strong seismic shaking can cause the release of submarine landslides in the form of *turbidity currents.* These are rapidly moving, sediment-laden slurries that slide down steep undersea canyon walls, pulled downward through the water by gravity. The evidence for scores of such turbidity currents have been found by retrieving samples from the ocean floor off the coasts of northern California, Oregon, and Washington. The timing of these currents—and the earthquakes they imply—shows a remarkable correlation with the timing of major earthquakes along the northern segment of the San Andreas Fault.

In the last 3,000 years, 14 of the 15 major earthquakes that occurred along the northern segment were preceded by a major earthquake along the Cascadia subduction zone. The average time interval between a Cascadia event and the subsequent San Andreas earthquake is 40 years, yet some of these paired events might have been simultaneous. The only major earthquake *not* preceded by a Cascadia event was the 1906 earthquake—which adds to the intrigue as to exactly why the 1906 earthquake was different.

But the important point is that the timing of major earthquakes along the San Andreas Fault seems to be influenced, at least in part, by factors that extend far beyond the main strand of the fault, over a region that extends north into the Cascadia subduction zone. In fact, to get a proper understanding of what causes earthquakes in California and how the San Andreas Fault is evolving, it is necessary to look not only across all of California, but across the entire American West.

CHAPTER 11

DISASSEMBLING CALIFORNIA

The whole place was shaking like crazy.

—A guest on the 18th floor of the Mirage Hotel,
Las Vegas, October 16, 1999

Two subtle clues to the history of the San Andreas Fault can be found at Weston Beach near Point Lobos south of Monterey Bay. One of the clues is the existence of distinct purple rocks with pink flecks that can be found embedded in a conglomerate at the south end of the beach.

A conglomerate, as every introductory geology textbook explains, is an easily recognized sedimentary deposit comprised of rounded boulders, cobbles, and pebbles set in a matrix of fine sand and silt. This particular conglomerate, part of the Carmelo Formation—named and first described by Lawson in 1892—formed as a result of an undersea landslide that slid down the side of a steep submarine canyon. It is the product of a turbidity current. Later, as the Coast Ranges were formed, the conglomerate was lifted up to sea level. How it was lifted does not concern us at the moment. Instead, focus on the purple rock.

I draw attention to this particular component of the conglomerate because it is easy to identify. About one out of every ten boulders, cobbles, or pebbles in the conglomerate is this purple rock peppered with pink flecks of feldspar crystals, which adds to its attractiveness and ease of identification. It is a volcanic rock—actually, part of a *volcanic tuff,* a thick ash deposit—exploded 150 million years ago from a volcano far to the south that has long been extinct and now likely eroded away. Keep in mind the existence of this purple rock at Weston Beach.

At the north end of Weston Beach is a light-gray sandstone with an abundance of fossil imprints. The imprints include mud cracks, animal burrows, and trails of animal tracks. All are easy to find and quite easy to see, especially when the angle of the sun is low and shadows are long in early morning or late afternoon. There is a peculiar imprint that looks like it was made by a feathered serpent, or perhaps a feather boa. Even experienced geologists have commonly mistaken it for some type of ancient seaweed, but it was part of an animal: It is the imprint of an inhalant siphon of the bivalve mollusk *Hillichnus lobosensis.*

The imprints of *Hillichnus lobosensis* in the sandstone at Weston Beach are one to two inches wide and as much as five feet long. They often occur in clusters, each individual imprint forming an arc. What makes it all the more intriguing is that *Hillichnus lobosensis* is a rare fossil, found at only two places along the California coast—at Weston Beach and at Point Reyes, 100 miles to the north.

At Point Reyes, there is also a light-gray sandstone that contains imprints of animal burrows and animal tracks and of *Hillichnus lobosensis.* There is also a conglomerate at Point Reyes with boulders, cobbles, and pebbles of a purple rock with pink flecks, best found just before making the descent to the lighthouse, hanging near the skull of the female gray whale that was put on display many years ago.

Hillichnus lobosensis and the purple rock are just two indicators that the rock sequences at Point Lobos and Point Reyes are similar;

there are more. In fact, every characteristic that has been studied—detailed structures in the sandstone that reveal the direction and speed of ancient ocean currents, microfossils in the conglomerate, other rock types in the conglomerate, as well as a white granite exposed as basement rock beneath the conglomerate at both Point Lobos and Point Reyes—points to the same conclusion: The rocks at Point Reyes and at Point Lobos were once adjacent. But how did they become separated by 100 miles?

The obvious answer is that they were slid apart by the San Andreas Fault, but an alert reader who knows California geography will realize that this is impossible because both Point Lobos and Weston Beach and Point Reyes and its lighthouse lie on the *same* side of the San Andreas Fault—on the west side.

So how do we account for this phenomenon? Perhaps there is *another* San Andreas–type fault west of the actual one, and it was movement along *this* fault that separated the rocks that are now at Point Reyes from those at Point Lobos. That, indeed, is what happened.

The San Gregorio-Hosgri Fault begins just south of Point Reyes and runs for more than 200 miles southward off the coast of central California, coming onshore at just a few places: at Bolinas Lagoon near Point Reyes; a short segment on the San Francisco Peninsula north of Santa Cruz; at Big Sur, where movement along this fault is responsible for the spectacular sea cliffs; and at San Simeon near Moonstone Beach not far from the famous Hearst Castle.

In every regard, the San Gregorio-Hosgri Fault can be considered an early version of the San Andreas Fault. It has slid blocks horizontally at least 100 miles; it was active earlier than the current San Andreas Fault; and it remains active today, though at a much reduced rate—the most recent earthquake of note was a magnitude-4.6 earthquake in 1963 in the Santa Cruz Mountains.

Today, as the Pacific and North American plates slide past each other in California, the motion is taken up mostly along the San Andreas Fault. Earlier, it was along the San Gregorio-Hosgri Fault.

It is one of the obvious clues that there has been an evolution of the Pacific–North American plate boundary, and the boundary continues to evolve, which explains why earthquakes occur across almost the entire state and why, according to the California Geological Survey, there are more than 700 different faults scattered across California that have ruptured in the last 10,000 years. California is, indeed, earthquake country.

This also provides a view as to the future of the state. California is not going to fall catastrophically into the ocean, as some doomsday predictors profess, but it is being sliced and slid apart incrementally, most of the sliding occurring during the occasional large earthquake.

This realization is not some inconsequential fact. Because earthquakes cannot be predicted—at least, not at present—it is not possible to pinpoint the time and place and exact severity of the next major one. But it is possible to know where the next potentially damaging earthquakes are most likely to occur. And this rests on knowing not only how the state of California is being torn apart, but how the various terranes that are California were assembled.

The East and the West Coasts of the United States are profoundly different socially, historically, and geologically. The East Coast is stable and urbane, while the West Coast is free-wheeling and sprawling and the people are highly mobile. And then there are the differences in geologic histories.

Long ago the East Coast was dominated by episodes of mountain building, called orogenies, interrupted by prolonged periods of extension and basin formation. The first such mountain-building episode, the Taconic Orogeny, occurred about 500 million years ago and was caused by a collision of an island arc, probably similar to the one formed by the Aleutian Islands today, against what was the nascent east continent of North America. The line along which the collision took place and where the island arc is held fixed today against the continent, a line appropriately known as a "suture," can be followed from Newfoundland to New Jersey. It passes through Maine, Massachusetts, and Connecticut. It runs along Long Island Sound and the Harlem River, then the East River, and crosses New York Harbor to Staten Island. If one cares to stand along it, an accessible spot can be found in the Bronx at Tremont Park, where a baseball diamond straddles it. The rocks exposed along the right-field line are those of the old continent—part of the Manhattan Schist—while those on the opposite, left-field side are the younger arrivals—deep-water shales deposited on oceanic crust—that once rode atop a tectonic plate and smashed against North America.

Another collision, the Arcadian Orogeny, occurred about 100,000,000 years later and involved the collision of two entire continents, forming the northern Appalachian Mountains. Then after a period when the two continents might have pulled apart, they were pushed together again, this time at a slightly different angle, and the Alleghenian Orogeny was the result, forming the southern Appalachian Mountains from Alabama to New Jersey.

Then the motion of the continents reversed and the continents were pulled apart again, but this time the pulling continued until a great rift formed a mid-ocean ridge and a new ocean crust was created. This was the breakup that formed the Atlantic Ocean. Since then, the East Coast of North America has no longer been a plate boundary where continents are shoved together and mountains created. Instead, it has been a passive margin and those mountains formed by the Acadian and Allegheian Orogenies, which were once as high as the Alps and the Rocky Mountains, have been eroded down to the sequence of rhythmic hills seen today.

Meanwhile, 500 million years ago when the Taconic Orogeny was happening on the East Coast, the West Coast of North America was a passive margin located about where the Nevada-Utah border is today. To the west was a vast ocean. Then about 350 million years ago, just after the Arcadian Orogeny, the West Coast changed in character—drastically.

Exactly how it happened is still debated, but the basic timeline is this: A mountain-building episode—this one known as the Antler Orogeny, perhaps involving a collision with an island arc—extended the West Coast as far as central Nevada. Then 50 million years later, a huge block of continental material of unknown origin—known as the Sonomia Block—collided with the West Coast and extended the coastline westward hundreds of miles into what is now eastern California. That was followed by the subduction of a series of oceanic plates. One of those oceanic plates carried a continental block—known as the Smartville Block—which arrived about 150 million years ago and pushed up next to the western edge of the Sonomia Block.* Oceanic plates continued to subduct under the continent of North America—the last one being the Farallon plate—and as that happened, parts of the ocean crust were scraped off to form the suite of rocks known as the Franciscan, such as the red chert visible on the north side of the Golden Gate, and, ultimately, to be the source of most of the colorful pebbles at Moonstone Beach. Thus, in piecemeal fashion, was the state of California assembled.

But there came a time when the western edge of the North American plate drifted far enough west to ride over the spreading centers separating the Farallon and Pacific plates, causing a sliver of the North American to detach itself from the rest of the continent and be captured and move with the Pacific plate. Ever since, a succession of transform faults has formed, the latest and currently the most active one being the San Andreas Fault.

---

* It is along the suture between the Smartville and the Sonomia Blocks that the Mother Lode, the stretch of major gold fields of California, is found.

The first transform fault developed along the western edge of the North American plate 25 million years ago in the broad offshore region of southern California where the Channel Islands stand today, a region commonly referred to among seismologists as "the borderlands."* These islands—San Clemente and Santa Catalina south of Los Angeles, and San Miguel, Santa Rosa, and Santa Cruz south of Santa Barbara—are all comprised of continental rocks and stand above sea level as the highest ridges of a complex basin-and-ridge area that lies off the coast of southern California. This offshore continental area is crisscrossed by a host of strike-slip faults. Which one might represent the ur–San Andreas Fault has yet to be determined. But this much is known: For the first few million years after the North American plate began to ride over the Pacific plate, whatever transform faults developed, developed in borderlands. There is no evidence of a San Andreas–type fault on land until about 20 million years ago.

These earliest known faults are no longer a single strand but have been cut into segments, which have been moved around by earthquakes that occurred along younger fault strands. But at least three segments of an early strand can still be found. One, known as the San Francisquito Fault, runs through the western end of the San Gabriel Mountains. Another, the Fenner Fault, also in the San Gabriel Mountains, lies a few miles east of Valyermo. And another, the Clemens Well Fault, lies far to the east in the Orocopia Mountains near the Salton Sea. The complex work of restoring geologic units to their original positions by undoing the sliding and rotation of blocks and straightening out folded and compressed sedimentary layers—a technique known as palinspastic reconstruction—shows that these three short fault segments once formed a single long strand. Furthermore, it shows that the single long strand was once the primary transform fault in California, that it was active until

* As mentioned in Chapter 8, the age of 25 million years comes from the oceanographic work by Atwater.

about 12 million years ago, and that 60 miles of horizontal displacement accumulated along it. But nothing is permanent, certainly not in the geology of California. And the San Francisquito-Fenner-Clemens Well Fault was eventually cut up and replaced by younger, longer strands as the primary movers and shakers in the region.

When the North American plate began to drift over the Farallon-Pacific's spreading central region, a transform fault formed, and then a peculiar feature developed at either end of that fault. The feature, known as a triple junction, is a place where the boundaries of three tectonic plates meet. In this case, two of the plates are the North American and Pacific plates; the third, which is actually what remains of the Farallon plate, has been given a different name depending on whether it is north or south of the transform fault. At the north end, the surviving part of the Farallon plate is now known as the Gorda plate and the point where the three plates meet is the Mendocino triple junction, because the point is currently located near Cape Mendocino. At the south end is the Cocos plate—a remnant of the Farallon plate—and the Rivera triple junction. What is important here is that, because of the directions in which the various plates are moving, neither the Mendocino nor the Rivera triple junction is stationary; both migrate. And they migrate in opposite directions, the Mendocino triple junction to the north and the Rivera to the south. As time progresses, the transform boundary between the Pacific and North American plates lengthens. And that brings us back to the San Francisquito-Fenner-Clemens Well Fault.

Twelve million years ago, when the San Francisquito-Fenner-Clemens Well Fault became inactive and a new fault strand formed, the northern triple junction was near San Francisco and the southern triple junction was somewhere south of San Diego. In northern California, the strike-slip movement between the Pacific and the North American plates was taken up by the San Gregorio-Hosgri Fault, which would slice through the conglomerate that contained the fossilized bivalve mollusk *Hillichnus lobosensis* and transport the western part northward to where it now resides at Point Reyes while the southern part remained at Point Lobos. In southern California, movement was primarily along the San Gabriel Fault, which runs along the western and southern boundaries of the San Gabriel Mountains. Then, after several million years, seismic activity shifted again in both northern and southern California.

In the north, activity shifted from the San Gregorio-Hosgri Fault eastward, perhaps first to the east side of what is now San Francisco Bay, then back to the west side, where it is today. In fact, the segment of the San Andreas Fault that runs through San Andreas Valley—the namesake of the fault and where Lawson first recognized the fault—may be the youngest strand of the entire fault, having formed one or two million years ago. It also has a relatively small amount of accumulated movement; in the San Andreas Valley, the fault has displaced rocks only about 20 miles.

If the position of the fault has shifted before, it will almost certainly shift again. A future candidate for taking up much of the motion between the Pacific and North American plates and replacing the current active trace of the San Andreas Fault is the Hayward Fault on the east side of San Francisco Bay. A look at a map of fault locations in the Bay Area shows that the Hayward Fault is nearly aligned with the Calaveras Fault to the south and the Rodgers Creek, Healdsburg, and Maacama Faults that continue north of San Francisco Bay. And so it seems natural that these individual fault strands might coalesce into a single long strand and become the primary fault strand that runs through northern California.

But the shift may not stop there. There is reason to believe that, millions of years from now, the primary fault strand between the Pacific and North American plates could shift hundreds of miles and be on the east side of the Sierra Nevada. To understand why this might happen, one has to understand how the southern segment of the San Andreas Fault has evolved.

Almost everything in southern California is of recent origin—including the geology.

The San Gabriel Fault was the main fault strand through southern California from 12 to about 4 million years ago and it may have extended farther east than it does today along what is known as the Cajon Valley Fault. Exactly what was the geologic layout before 4,000,000 years ago has been difficult to determine because almost every square foot of land in southern California has since been disturbed. Almost every mountain, every hill, every river, every canyon, and every fault has either been shoved away from its original location or disrupted or come into existence in the last 4,000,000 years. If one wanted to attribute this regional chaos to one entity, it would be the erratic migration of the Rivera triple junction at the southern tip of the San Andreas Fault.

The Mendocino triple junction at the north end of the San Andreas Fault has had a simple history: It has migrated northward at a steady rate, leaving behind, as evidence of its migration, a progression of volcanic activity. South of San Jose are the Quien Sabe volcanics in the Diablo Range, which erupted about 12,000,000 years ago. North of San Jose, in the Berkeley Hills east of Oakland, are the Moraga volcanics, dated at 10,000,000 years. Near Calistoga, north of San Francisco Bay, are the Sonoma volcanics, where the heavy fall of volcanic ash produced a petrified forest of redwood trunks 6,000,000 years ago. And farther north, 80 miles south of the Mendocino triple junction, is the Geysers geothermal field, where the most recent volcanic activity was a mere 10,000 years ago. The Rivera triple junction has not left such a trail of volcanics—nor has its migration been as simple.

In fact, there may have been multiple triple junctions at the southern end of the developing transform fault that eventually coalesced into one. This much is certain: About 4,000,000 years ago, what is today known as the Rivera triple junction jumped a few hundred miles east to a point under the North American plate, shearing off a long slab of continental material that became the peninsula of Baja California and forming the Gulf of California.

The Gulf of California formed because of the development of a series of short spreading centers connected by a series of equally short transform faults, creating new ocean crust. After 4,000,000 years, the opposite sides of the gulf have spread apart 200 miles. And the seaway would extend all the way north as far as Indio, California, except for a surprising geologic consequence—the creation of the Grand Canyon.

As the Gulf of California spread open, the crust thinned and the land surface dropped. In California, this produced a long broad feature known as the Salton Trough, which runs as far north as Indio. Studies show that this trough should be as much as three miles deep, except that it is filled with river sediments. And it is filled with river sediments because the formation of the Gulf of California and the lowering of the land surface steepened the gradient of the Colorado River.

The steepening of the gradient led to a brief period of rapid erosion and the formation of the Grand Canyon. Much of what was removed now fills the Salton Trough. And then, there is a modern twist to the story.

When western settlers first arrived in California, the Salton Trough was a desert, barely able to support farming. But in 1900, a canal system was constructed to bring water from the Colorado River to the Salton Trough. To attract more farmers, the area was advertised as the Imperial Valley. The canals were indeed successful and brought the needed water, but they soon silted up, so more canals were constructed. The construction of the second round of canals was streamlined by bypassing control gates. Then a series of floods occurred, so that by October 1905, virtually the entire Colorado River was flowing into the Salton Trough. The flow of river water was finally stopped in 1907, but by then a saline lake had formed: the Salton Sea.

The twist is that, in the past, water from the Colorado River has been diverted naturally into the Salton Trough, forming a large lake. At those times, paleoseismological studies show that earthquakes along the San Andreas were frequent; that is, the weight of the lake water *may* have modulated the earthquakes. If so, the question can be asked: Would the refilling of the Salton Trough trigger an earthquake along this segment of the San Andreas, which has not had a major earthquake since 1690? Moreover, could this paradoxically be a way to control earthquakes?

The filling of water reservoirs behind dams has triggered some major earthquakes, the most serious case being the magnitude-7.9 Sichuan earthquake in south-central China in May 2008 that killed 80,000 people. This *might* have been avoided if we knew more about how the weight of water in a reservoir actually triggered the earthquake and whether a partial filling of a reservoir could insure the occurrence of a series of moderate earthquakes and not a single large destructive one. And so I return to consider how the formation of the Salton Trough and the northwest drift of Baja California have affected the evolution of the San Andreas Fault.

For 4,000,000 years, after its separation from North America, Baja California has been ramming into southeastern California. The net result is that the greater Los Angeles area is caught in a vise. The area is being squeezed, and that has caused the crust to buckle and throw up an east-west line of mountains—the Transverse Ranges, which run from Santa Barbara to San Bernardino—and is responsible for the "Gordian knot" of geology declared by Josiah Whitney when he began to examine California geology.

At the same time, the Los Angeles area is also being dragged to the northwest by the Pacific plate. All of this tectonic action has created a complex network of active faults that, when plotted on a map, looks like a shattered pane of glass. And along the northern and the eastern edges of the pane is the San Andreas Fault.

The Mojave segment of the San Andreas Fault runs along the northern edge of the San Gabriel Mountains. This segment is remarkably straight and came into existence when the mountains did, about 4,000,000 years ago—replacing the old San Gabriel

Fault with the modern San Andreas strand. But to the east, after it passes through Cajon Pass, the modern San Andreas Fault loses its straight-line simplicity, and even today geologists are unsure exactly where the San Andreas Fault runs.

The general opinion is that it separates into two strands—the Mission Creek and the Banning Faults—that run along the north side of San Gorgonio Pass, which connects the greater Los Angeles basin and the Salton Trough. The southern strand—the Banning Fault—passes a few miles east of the luxury resorts and celebrity mansions at Palm Springs, then reconnects with the Mission Creek Fault at the southern end of the Indio Hills. From there, the San Andreas is again a single straight strand and is easy to follow another 40 miles to its end at the east edge of the Salton Sea.

But it is the complexity between Cajon Pass and Palm Springs that has geologists pointing to that segment of the San Andreas Fault as going through a rapid evolution.

If, instead of following the evidence of surface faulting that leads one to the Mission Creek and Banning Faults and through San Gorgonio Pass, one looks at a map of seismicity and follows the most prominent line of recent earthquakes, one would pass through the city of San Bernardino, south to Colton, then through Moreno Valley, over the San Jacinto Mountains, down through Borrego Valley, and end at the Superstition Hills on the *west* side of the Salton Sea. This is the San Jacinto Fault and it is the *most* seismically active fault in southern California, meaning it has had more large historic earthquakes than any fault—including the San Andreas—in southern California.*

To add another level of complexity—and intrigue—there is yet another well-developed, horizontally slipping fault in southern California. The Elsinore Fault runs between the San Jacinto Fault

* To further add to the recognition of the San Jacinto Fault, there is a four-tiered freeway interchange at Colton—where I-215 and I-10 meet—that is built directly over the fault.

and the coast. Furthermore, when the southern segment of the San Andreas Fault is considered, there is a progression in age and activity among the three faults: The Elsinore Fault is the oldest and least active; the southern segment of the San Andreas that runs through San Gregorio Pass and close to Palm Springs is middle in age and in activity; and the San Jacinto is the most active and the youngest, probably forming in the last 1,000,000 or 2,000,000 years.

So the continual opening of the Gulf of California has had the following effect: It caused a transfer of activity from the Elsinore Fault to the San Andreas Fault, then from the San Andreas Fault to the San Jacinto Fault.

At first glance this seems of no consequence, until one considers it on a large scale—a continent-wide scale—and tries to understand exactly how the western edge of the North American plate is being disrupted by the northwest motion of the Pacific plate. And to do this, one finds the final crucial element by leaving the hot, arid land of Los Angeles and the Salton Trough and traveling 1,000 miles north to the cool Ponderosa pines at the edge of Lake Tahoe. At that point, one is standing at the northern end of another trail of earthquakes—the little-known Walker Lane seismic zone.

Located along the border between California and Nevada, Lake Tahoe is the largest alpine lake in North America and the second-deepest lake in the United States after Crater Lake in Oregon, which was formed by the collapse of Mount Mazama during a volcanic eruption 8,000 years ago. The formation of Lake Tahoe was decidedly different.

Lake Tahoe sits in a basin that formed in two stages. Beginning about 2,000,000 years ago, it grew as a result of an east-west pull that produced the Basin and Range province, which covers most of Nevada and that created the high peaks of the Carson Range on the east side and the Sierra Nevada Mountains on the west side of

Lake Tahoe. Then, about 1,000,000 years ago, a set of horizontally slipping faults began to develop in a broad zone that runs along the California-Nevada border.

The zone begins somewhere north of Lake Tahoe—perhaps, as some geologists say, at Pyramid Lake north of Reno, which is also in a basin, or, as seismologists who locate earthquakes argue, as far north as south-central Oregon. In any case, the zone runs south of Lake Tahoe and includes Owens Valley in California and the Yucca Mountains in Nevada—where the infamous Area 51 is located, the source of many conspiracy theories about captured UFOs, but which in actuality is a supersecret test site for military aircraft.

The zone is a place of persistent earthquakes known as the Walker Lane seismic zone. It was named for Joseph Walker, leader of a military expedition that in 1833 found what became known as the California trail, the primary route used by immigrants to reach the gold fields, and, that same year, probably the first easterner to gaze upon Yosemite Valley.

Though the seismic activity along the Walker Lane is less than along the San Andreas and its associated faults, it is not insignificant. The most recent flurry of activity near Lake Tahoe occurred in April 2008 when more than 600 earthquakes shook the region, the largest a magnitude-4.7 event on April 15. The last significant earthquake was a magnitude-6.1 tremor in 1914.

The largest historical earthquake in the Walker zone was the 1872 Owens Valley earthquake, studied by Josiah Whitney and Grove Karl Gilbert and felt by Muir in Yosemite Valley. As Gilbert noted, there was a large amount of horizontal displacement in the same direction as slip for large earthquakes along the San Andreas Fault—that is, the Sierra Nevada Mountains moved northward with respect to the floor of Owens Valley and the rest of North America. And that is not a coincidence. It shows that the effect of the northward motion of the Pacific plate is being felt at least as far east as the Walker Lane seismic zone, and in some way is responsible for the formation of the basin that holds Lake Tahoe. But that is not the limit of the effect of the Pacific plate on North America, or at least that is not the limit of whatever forces inside the Earth are driving the motion of the Pacific plate.

A snapshot of the speed and the direction that various points in North America are moving can be determined quite precisely by use of the Global Positioning System, the familiar satellite-based system used to navigate cars or, with a handheld electronic device, to find the nearest Italian restaurant or specialty boutique. To determine precise movements of the Earth's surface, highly sophisticated equipment is used and complex data-handling procedures are applied. The results are impressive because after a year or so of measurements, speeds as low as a few tenths of an inch per year can be determined for points separated by thousands of miles.

Such measurements show that the Pacific plate is moving to the northwest at a steady rate of 1.9 inches per year relative to the interior of North America. Two-thirds of the movement—1.4 inches per year—is occurring across the San Andreas Fault and its subsidiary faults, such as the San Gregorio-Hosgri, Hayward, and San Jacinto Faults. A quarter of the relative plate motion—0.4 inches per year—is occurring across the Walker Lane seismic zone, which is consistent with the lower seismic activity of the Walker Lane compared to the San Andreas system. But that still leaves a small amount of plate motion—0.1 inches per year—to be explained.

And GPS measurements have revealed this motion: It is occurring across yet another seismic zone—the Intermountain Seismic Belt—that bisects Utah from south to north, includes the Wasatch Fault east of the Great Salt Lake, which broke in at least five distinct major earthquakes between about 400 and 1600 A.D., and that may run as far north as Idaho and Montana. Major historical earthquakes have occurred along this seismic belt, most recently as a magnitude-7.3 earthquake near Hebgen Lake, Montana, in 1959 and a magnitude-6.9 earthquake near Borah Peak, Idaho, in 1983.

Furthermore, the broad regions between these various seismic areas—the triangular block comprising of the Great Valley of

California and the mountainous Sierra Nevada that is bounded on the west by the San Andreas Fault system and on the east by the Walker Lane, and the elongated block comprising most of the state of Nevada and that is situated between the Walker Lane and the Intermountain Seismic Belt—are, essentially, aseismic and are moving as rigid blocks. And that, to sum it all up, is what is happening now to the western part of the United States today.

As described earlier in the chapter, for more than 100,000,000 years huge crustal blocks, such as the Sonomia and Smartville Blocks, to name two, collided against North America to form much of the western United States. That changed 25 million years ago when the earliest San Andreas–type fault formed. Since then, increasingly large amounts of the western United States are being sliced apart—the Los Angeles borderlands and basin, the Great Valley-Sierra Nevada, and central Nevada—and these are being slid to the northwest by the same internal force that drives the Pacific plate.

So a very long and very grand moment is being played out in the western United States. The San Andreas Fault is the primary element today in the boundary between the Pacific and North American plates, taking up almost all of the relative plate motion and accommodating it, mostly, by the occasional large earthquake. But it is not the only element. Others include the Walker Lane, where a new San Andreas–type fault might be in the making, and, farther east, the Intermountain Seismic Belt, which in the distant future will become the primary element in the plate boundary.

Whether this will come to pass is, of course, unknown. But if current trends continue for another 10 or 20 million years—and hundreds of thousands of major earthquakes occur—the Salinian block west and south of San Francisco, the Los Angeles borderlands and basin, and the peninsula of Baja California will be several hundred miles northwest of their current positions and separate from the North American continent. By then, the ever-growing Gulf of California might reach into Owens Valley, and Las Vegas could be near the Pacific shore. And *maybe* a continuous right-lateral, strike-slip San Andreas–type fault might run close to Lake Tahoe and through northern California and part of southern Oregon.

But there is an important element missing in this story: the opening of the Gulf of California that caused an eastward shift of seismic activity and moved the southern segment of the San Andreas Fault to its current location. The 1872 Owens Valley earthquake supports the idea that the opening of the Gulf of California is also responsible for the creation of the Walker Lane seismic zone. But what is happening in between? What is happening across the Mojave Desert?

For years, there was speculation as to what *must* be happening across the Mojave Desert to support the above story, but nothing was clear until June 1992, when the largest earthquake to strike southern California in 40 years hit. That earthquake started near the San Andreas Fault, then ruptured northward across 60 miles of the desolate Mojave Desert, opening a new chapter in understanding how the San Andreas is evolving.

Undoubtedly, the most important earthquake so far in California history was the 1906 San Francisco earthquake because of the devastation it caused to the city and because of the advancements that followed to the new science of seismology. Arguably, the second most important earthquake is one that did almost no damage and caused only one fatality—that of an unfortunate three-year-old boy who was asleep when a cinder block from a chimney crashed through his house. It is this second important earthquake that seismologists have focused on because it showed what seismologists had long expected—that earthquakes, even those separated by many years, interact.

On June 28, 1992, early morning gamblers in Las Vegas paused for the few seconds it took seismic waves to roll under them.

In Hollywood, a man who was awakened by the early morning shaking said it felt, for a full 30 seconds, as if his house were a ship on a rough sea. Thirty miles north of Palm Springs in the town of Landers, where the earthquake originated, a Mrs. Iona Bong, 61 years old, said she started to fly off her bed when, fortunately, her husband grabbed her arm and brought her back to earth. The retired couple spent the next day sweeping up broken glass and smashed belongings and carting them to a huge garbage can. For the seismologists who would study the 1992 Landers earthquake, it was, as one seismologist pegged it, "a revolutionary event."

First, the earthquake did not occur on one, but on *six* different faults. The rupture began on a short segment of a previously unknown fault located about 20 miles north of the San Andreas Fault and the northern end of the Salton Trough. From there, it propagated northward along the Johnson Valley Fault, then onto the Landers Fault. After a pause of seconds—which can be read on the seismographs recorded during the earthquake—the rupture resumed and continued to break northward along three other nearby faults in succession: the entire length of the Homestead Fault, the northern half of the Emerson Fault, and the southern third of the Camp Rock Fault. In all, the surface rupture extended for 60 miles across the Mojave Desert along a segment of a line that could be drawn from Lake Tahoe, down the axis of Owens Valley, and to the Salton Trough. Here was confirmation that the system of valleys, such as Owens Valley, which were contained within the Walker Lane seismic zone should be considered a northern extension of the same tectonic forces that were opening the Gulf of California.

The intermediate seismic zone that crosses the Mojave Desert has since been known as the Eastern California Shear Zone. Earthquakes in that zone cause the land to slip to the right, just like those of the Walker Lane and the San Andreas. In short, all three areas are involved in accommodating the motion between the Pacific and the North American plates, and so all three are part of the plate boundary.

Then there is the question as to why so many different faults were involved in a single event. Why did so many individual earthquakes happen on so many nearby faults?

Long before the 1992 Landers earthquake, seismologists had speculated that an earthquake on one fault might trigger earthquakes on other nearby faults. Here was ample proof: The six faults had broken in quick succession, the whole process—again, determined by a careful examination of seismic records—taking about 25 seconds. And that was followed by the usual occurrence of aftershocks, a swarm of earthquakes smaller than the main shock that originate in the same area.

Then came a surprise.

Three hours and eight minutes after the Landers earthquake, another earthquake rocked southern California. It originated under the San Bernardino Mountains, 30 miles west of the town of Landers near Big Bear ski resort. It was outside the expected area of aftershocks, so seismologists were at first unsure how to classify it. Was it a distant aftershock? Was it an independent earthquake, and the two events had occurred within hours of each other by happenstance? Or had the Landers earthquake triggered, in some yet to be explained way, a significant earthquake beneath Big Bear?

Seven years later, another major earthquake struck southern California, and again it originated beneath the Mojave Desert. This one was in an even more remote location than the Landers earthquake; it was centered beneath an abandoned rock quarry, and it is now known as the Hector Mine earthquake, a magnitude-7.1 event that occurred on October 16, 1999.

Candy McCain of Phoenix, who was feeding coins into a slot machine in Las Vegas, felt the shaking. "I wasn't sure exactly what was going on," she told a news reporter, "but then I saw the signs swaying and the leaves on the fake palms rustling, so I thought it must be an earthquake. When it stopped, I started playing again."

Michelle Fabian, who was in bed asleep on the 18th floor of the Mirage Hotel, was awakened by the shaking. "The whole place was shaking like crazy," she said.

A westbound Amtrak train, the Southwest Chief, was traveling near the earthquake's center when the shaking started. Fortunately, it was behind a freight train and was going slower than usual. But the combination of the shaking and the moving train caused several rails to come loose and the Southwest Chief derailed, injuring several people.

Though the damage was slight and the injuries minor, the Hector Mine earthquake is important because it is only the second time in California history—that is, during the last century and a half—that a major earthquake has occurred beneath the Mojave Desert. It and the Landers earthquake—the other major earthquake to occur beneath the desert—happened within just seven years of each other, an unlikely coincidence when one considers that paleoseismic work showed that neither the Landers nor the Hector Mine Fault had slipped in more than 1,000 years.

So something extraordinary happened in the Mojave Desert in the 1990s. Somehow, the 1992 Landers earthquake triggered the Hector Mine earthquake seven years later. It seemed the theory of elastic rebound that Reid proposed after the 1906 earthquake had to be modified and applied to *multiple* earthquakes. In short, it seemed that large earthquakes should not be considered single isolated events, but part of a sequence or cluster, a conclusion that Sieh and others had already been led to in their paleoseismological work.

As scientific coincidences go, a year after the Landers earthquake and a few years before the Hector Mine earthquake, an expert on the mechanics of earthquakes was in Greece for a scientific meeting. During that meeting, he happened to see something in the archaeological record that convinced him that, yes, indeed, major earthquakes could occur as quick sequences, one triggering the other. And he gave a name to the new phenomenon: He called it an "earthquake storm."

CHAPTER 12

EARTHQUAKE STORMS

When you get a lot of earthquakes, you get *a lot* of earthquakes.
—Charles Richter

It is often hard to see the obvious, unless you are someone who brings a new perspective to a problem. Stanford professor Amos Nur, an expert on earthquakes and in the more general field of how rocks fracture when subjected to high pressure, saw something at the ancient Greek city of Mycenae that countless others had simply overlooked.

It was 1993 and Nur was attending a conference on archaeo-seismology, a relatively new term to describe how archaeologists and seismologists were trying to join forces and benefit from each other's work. But as Nur later recounted, the conference was a disappointment. The archaeologists and the seismologists seldom mixed, except during short breaks when both groups would indulge in drinking strong Greek coffee, and during an occasional day trip when they were taken to see one of the nearby ancient sites. It was during one such trip that Nur and the others visited Mycenae.

Here a quick diversion explains an unexpected circumstance that links Mycenae to California and its geology. In 1851, Heinrich Schliemann, then a German businessman, made a trip to California to claim his dead brother's estate. His brother had been one of the first to arrive in the gold fields, though instead of searching for gold, the brother made a quick fortune buying and selling claims. When Schliemann arrived, he too sensed an opportunity and opened a bank in Sacramento, where he traded in gold dust. The venture was short-lived; local agents were soon complaining that they were receiving short-weight consignments, and Schliemann, feigning illness, quickly left. He eventually made his way to the eastern Mediterranean, where he used his brother's fortune—and whatever additional money he had accumulated while in California—to finance archaeological work at Mycenae and Troy and other soon-to-be-famous sites. More than a century later, Nur, standing at Mycenae, recognized a feature that Schliemann had unearthed that would change the way seismologists determine earthquake risk in California.

What Nur saw is along the entranceway to the ancient city and within sight of the famous Lion Gate, where a stone relief depicts two lionesses in upright heraldic positions. It was through this gate, so tradition says, that Agamemnon, a Mycenaean king and one of the main characters in Homer's *Iliad,* marched his army and led them on a ten-year siege of Troy. Just outside the gate is an immense stone wall that sits atop a head-high steep incline of highly polished rock. To most people, the incline conveys a sense of rock-solid security. To Nur, it was evidence of a past calamity.

Nur recognized the rock incline as a fault scarp—a line along which a past earthquake had fractured and thrust the ground upward. That meant the Mycenaeans had built their city over an active fault. Fortunately for them, this particular fault has not moved in thousands of years, but earthquakes are common in the region. Nur realized that the ancient city of Mycenae, which was at its greatest influence during the Bronze Age, must have been subjected to repeated seismic shakings. But what effect might such subterranean activity have had on the city's history? Might the sudden abandonment of Mycenae around 1200 B.C. have been caused by an earthquake?

Archaeologists said no. Though many major cities in the eastern Mediterranean—including Thebes in Greece, Knossos on Crete, and Troy in western Turkey—were also destroyed around 1200 B.C., and though every major site—from Pylos on the Peloponnese Peninsula, to Aleppo in Syria, and Ashkelon in southern Israel—shows some damage consistent with seismic shaking happening around 1200 B.C., the widespread destruction that brought about the end of the Bronze Age—"the worst disaster in ancient history, even more calamitous than the collapse of the Western Roman Empire," according to one noted classicist—was not instantaneous but had occurred over several decades.

For that reason, archaeologists argued that the end of the Bronze Age was probably caused by several factors, including invasions by foreign peoples and by internal political strife. But Nur proposed another idea.

Since a *single* catastrophic earthquake could not have been the cause, Nur suggested that *several* major earthquakes had struck a broad region of the eastern Mediterranean over a period of several decades. But was there any evidence that such a sequence of major earthquakes anywhere in the world had occurred in quick succession? Nur pored through catalogues of ancient earthquakes and discovered that, indeed, there was.

One such period of increased seismic activity had started in A.D. 343, when an earthquake struck northeast Turkey. Then in A.D. 358, a second earthquake happened to the west. Others followed, also in northern Turkey, in 362 and 368. Then between 394 and 412 A.D., six earthquakes occurred near Constantinople, modern-day Istanbul. Also during the late fourth and early fifth centuries, major earthquakes shook the southern Italian peninsula, the island of Sicily, and Libya in northern Africa, as well as the Holy land and

Cyprus. One of the largest events occurred in 365, when the southern shoreline of Crete was pushed up as much as 27 feet, comparable to the maximum amount of uplift recorded along the coast of Alaska in 1964—meaning the 365 earthquake had been a colossal event. In all, during the second half of the fourth century and the first few decades of the fifth century A.D., at least a dozen damaging earthquakes hit the central and eastern Mediterranean region. Nur's examination of earthquake catalogues also showed that the centuries immediately before and after were periods of relative seismic quiet.

But was there a more recent—a more obvious—example of a series of major earthquakes that had detailed information about the location and size of individual events? Yes, there was.

In 1939, after two centuries of quiescence, the North Anatolian Fault, which runs across northern Turkey roughly parallel to the coastline of the Black Sea and which is a boundary between the African and Eurasian plates, came to life. By 1999, 13 major earthquakes had occurred. What is even more remarkable, 7 of the 13 ruptured the North Anatolian Fault in a systematic way: Each successive earthquake ruptured a segment of the fault that was immediately west of the previous earthquake.

The sequence began in northeast Turkey—as it had in A.D. 343—near the city of Erzincan, where, on December 26, 1939, the shaking was so severe and the damage so great that the old part of Erzincan was abandoned and a new city center was soon built to the north. Then three years later, in 1942, the next earthquake happened, immediately west of Erzincan, and a year later yet another earthquake west of the 1942 event. In all, the sequence of seven west-migrating earthquakes ruptured a 600-mile-long continuous segment of the North Anatolian Fault.

Each of these prolonged releases of seismic energy—in the fourth and fifth centuries A.D. and again in the 20th century—lasted several decades and occurred after a millennium of relative seismic quiescence. This supported Nur's suggestion that a similar series of major earthquakes could have occurred over several decades around 1200 B.C., nearly a millennium before the recorded quakes of the fourth and fifth centuries. It was a new phenomenon, one that Nur called "an earthquake storm."

But why would seismic energy be released as a series of large earthquakes lasting for decades?

Nur had an answer: stress transfer.

I once spent an intriguing afternoon watching an artisan prepare colored glass panes for a stained-glass window. The secret, he revealed, was not to cut completely through a thick pane, which could shatter the glass into shards, but to etch each one with a cutting tool that left the geometric curve he wanted the edge of a pane to have. Then, by the appropriate application of heat and cold, by twisting the pane ever so slightly, and by relying on the weakness of an etched curve, he could induce a glass pane to break as a series of arcuate cracks and produce a pane of any desired shape.

The purpose of the application of heat and cold and of the twisting was to induce a specific pattern of concentrated stress that enabled the artisan to control where the pane was mostly likely to break.

It is an art that few people have ever mastered. If one substituted the induced thermal and twisting stresses for the buildup of stress in the earth's crust by the movement of tectonic plates, and substituted the sequence of cracks produced in the glass panes for earthquakes, then one can understand how an earthquake storm could be produced.

Think of it in another way. Imagine that a giant zipper is holding together two tectonic plates. As the two plates tug against each other, a segment of the zipper sudden slides open, but, as a zipper is apt to do, it snags occasionally. As the tugging continues, the zipper again slides, then snags again. Each time, the sliding zipper represents an earthquake and the tugging of the plates becomes concentrated at another place along the zipper.

Or consider another example—one that Nur prefers. Take a wide rubber band and cut a few short slits in it. As the band is stretched, each slit in turn opens up and the ends of the slits lengthen. The sequence that the slits open and by how much depends on how the stress pattern gets transferred and concentrated at new locations across the rubber band.

If this seems complicated, rest assured it can all be explained mathematically by applying what is known as the Coulomb-Navier failure criterion, a well-established physical law widely used by engineers to design buildings, bridges, and other monumental structures. The Coulomb-Navier failure criterion tells how much an object—or the Earth's crust—can be pushed or pulled, twisted or sheared, before it breaks. And the criterion has probably never been more thankfully applied—at least in a geologic application—than during the recent earthquake storm along the North Anatolian Fault in northern Turkey.

In 1997, using the Coulomb-Navier failure criterion, a forecast was made, based on the sequence of recent earthquake ruptures, that there was a 12% chance that a magnitude-7 or larger earthquake would strike near the city of Izmit, 40 miles east of Istanbul, during the next 30 years. Two years later, a magnitude-7.6 earthquake did devastate Izmit, killing more than 25,000 people and causing $65 billion in damages. Within months after that earthquake, another forecast was made, this time for the area around Düzce, 60 miles east of Izmit. Some school buildings, thought to be in danger of collapse by seismic shaking, were closed. Then on November 17, 1999, another earthquake hit, flattening school buildings.*

Such success gives credibility to earthquake forecasting, or rather to the idea that the probability of a future earthquake—identifying

* As often happens in science, several people may have the same idea at about the same time. In this case, the idea that stress transfer could trigger a series of major earthquakes was proposed by several scientists. The successful forecast made in 1997 for additional earthquakes in north Turkey was made by a trio of researchers—Ross Stein, Aykut Barka, and James Dieterich.

the magnitude and a time period—can be given based on stress transfer. It also prompted a search for other examples of earthquake storms throughout the Earth's history.

An earthquake storm probably ran up and down the Italian peninsula during the late 17th and throughout most of the 18th centuries. It began with two damaging earthquakes that originated beneath the Apennine Mountains east of Naples in 1694 and 1702. The activity then migrated north to a region east of Rome with three major earthquakes in early 1703. By the second half of the 18th century, activity had returned to southern Italy, where *five* major shakings occurred along the toe of the Italian boot, in Reggio Calabria.

A more recent storm occurred in eastern Mongolia between 1905 and 1957, when *four* magnitude-8 events struck. And an earthquake storm is happening now along the Xianshuihe and adjacent faults along the northern edge of the Tibetan plateau in southwest China, where 11 major earthquakes have happened in the last 120 years.

After decades of almost no seismic activity, the Xianshuihe Fault became active in 1893 when an earthquake rocked the Tibetan district of Kada, destroying the Dalai Lama's Grand Monastery of Hueiyuan and seven smaller monasteries. In all, 74 Buddhist priests and 137 Chinese and Tibetan soldiers were killed. Since then, ten more strong shakings have occurred, including a magnitude-8.0 shock on the nearby Longmenshan Fault in 2008. The most recent event occurred on April 14, 2010, in Qinghai Province when many Buddhists were killed when a 12th-century monastery collapsed.

This, of course, raises the question: Has an earthquake storm ever occurred in California? Here we are hampered by a historical record that spans barely 200 years. But using the techniques of paleoseismology, evidence has been revealed that *two* storms may have occurred in a place that few people associate with devastating earthquakes—Hollywood.

Though it is one of the most densely populated regions of California, Hollywood offers an unusual opportunity to recognize and walk along an active fault. The area was urbanized in the 1920s, before the widespread use of mechanized earth-moving equipment, so much of the original topography is still intact, even subtle features such as alignments of low hills and shallow troughs that record the trace of recent earthquakes. In essence, the network of winding streets and the placement at odd angles of apartment buildings and commercial enterprises, as well as the occasional abrupt slope across one of the sprawling lawns in Hollywood and nearby Beverly Hills, are subtle evidence of an original jumbled ground surface. And by finding the appropriate steep incline, one can follow the Hollywood Fault.

Begin at the corner of Hollywood and Vine and look north along Vine Street beyond the 13-storied cylindrical tower that houses Capitol Records. Just beyond Capitol Records, just before Vine Street reaches the Hollywood Freeway, the roadway ramps up a steep hill. The hill is there because the ground was pushed up by repeated earthquakes. Along the base of the hill is the Hollywood Fault.

From that point, the fault can be followed west along the base of the same hill, running parallel to and maintaining a distance of a few blocks north of Hollywood Boulevard. It runs along the base of the low hill where the Magic Castle, a private club of magicians and the home of the Academy of Magical Arts, is located. Farther west, the fault runs directly beneath the house where the Nelson family lived and where the opening scene of their famous 1960s sitcom—the series was called *The Adventures of Ozzie and Harriet*—was filmed.

Continuing west, close to where Hollywood Boulevard ends, the fault angles to the southwest and crosses just south of the busy intersection of Sunset and La Cienega Boulevards. This is a neighborhood of fine restaurants and fashionable boutiques. On the north side of Sunset Boulevard one can find, after considerable searching through the urban construction, an occasional outcrop of hard granite. This is the rock that comprises the Santa Monica Mountains to the north; high up the mountainside is the famous HOLLYWOOD sign. South of Sunset Boulevard, there are no rocky outcrops; instead, one stands on a deep layer, several hundred feet thick, of loose sediments that washed out of the canyons of the Santa Monica Mountains and that fill the Hollywood Basin. It is this discontinuity—granite outcrops north of Sunset Boulevard and deep sedimentary fill to the south—that, here, defines the Hollywood Fault.

The western end of the fault lies somewhere near the grounds of the Beverly Hills Hotel, just north of the intersection of Sunset Boulevard and Rodeo Drive. From there, if one walks a mile or so south along Rodeo Drive to Santa Monica Boulevard, then turns right and continues to Wilshire Boulevard, one will now be standing at the eastern end of another fault—the Santa Monica Fault—which continues to the ocean's edge and beyond.

Now return to where the Hollywood Fault crosses under Vine Street and head east. From here, the fault runs close to Franklin Avenue, then along Los Feliz Boulevard. At the east end of the Santa Monica Mountains—that is, at the southeast corner of Griffith Park, home of the Los Angeles Zoo and Griffith Observatory—the fault disappears under the floodplain of the Los Angeles River. What lies on the other side?

There is another fault—the Raymond Fault—which is, perhaps, a continuation of the Hollywood Fault and which runs eastward through southern Glendale and across the San Gabriel Valley, through South Pasadena to the foot of the San Gabriel Mountains. Kinks in the Raymond fault are responsible for the low hill where the luxurious Langham Hotel—formerly Ritz-Carlton—is perched and for the shallow depression that is Lacy Park. The Raymond Fault is also responsible for the low hills on the north side of the Santa Anita Racetrack, visible from the grandstand.

What do the Santa Monica, Hollywood, and Raymond Faults have in common? Besides lying along what seems to be a continuous line, all three ruptured about 10,000 years ago and again about 1,000 years ago.

Unfortunately, paleoseismologists have not yet determined whether the earthquakes along the Santa Monica, Hollywood, and Raymond Faults occurred as a single colossal event or as a series of relatively quick earthquakes, happening over years to centuries, the latter being an earthquake storm. (Unfortunately, the techniques used in paleoseismology are not yet sufficiently refined to distinguish, in this case, between years and centuries.) But there is a curious coincidence: All three faults did rupture at about the same time; then, after a period of several thousand years, all three ruptured again, lending further credence to Richter's statement: "When you get a lot of earthquakes, you get a *lot* of earthquakes."

Moreover, other nearby faults have a similar history.

James Dolan at the University of Southern California has dug trenches and sunk holes large enough for him to climb down to examine the Hollywood Fault. He has also dug trenches and sunk holes into the Puente Hills Fault that runs southeast from Griffith Park, the Whittier Fault that runs east of downtown Los Angeles, and the Newport-Inglewood Fault that runs south from close to the Beverly Hills Hotel to the city of Long Beach and may merge with the Rose Canyon Fault that continues to San Diego. At all of these faults, Dolan has determined a similar rupture history: major earthquakes along each one about 10,000 years ago and again about 1,000 years ago. And in each case, the earthquakes that ruptured these faults were much larger than the most recent damaging earthquake to strike the Los Angeles area, the 1994 Northridge earthquake, which killed 60 people, injured more than 7,000, and caused $44 billion in damages. In short, according to Dolan, there have been two "bursts" of seismic activity in Los Angeles and the immediate surroundings in the last 10,000 years.

Fortunately, the time interval between such "bursts," or earthquake storms, in this particular region of California is thousands of years, so it is highly unlikely that one will occur in the near future; thus this region is in a "seismic lull." But that is not true elsewhere in California.*

In 2008, a report was issued by the Working Group on California Earthquake Probabilities—a group that then consisted of about 50 geologists, geodesists, and seismologists—that said it was "virtually assured" that California would be struck by a magnitude-6.7 or larger earthquake during the next 30 years. Such a claim is not profound when one considers that a dozen such events occurred somewhere in California during the previous 100 years. What was profound was that the group was able to identify which faults were the most likely to rupture. The 2008 report has now been updated to indicate how severe the ground shaking might be and the chance of multiple earthquakes.

In northern California, the most probable destructive seismic event is along the Hayward Fault, which runs along the east side of San Francisco Bay. The previous event, in 1868, occurred when only 24,000 people lived near the fault. Today more than 1,000,000 people live within five miles of the fault trace. Hundreds of homes and other structures are built close to the fault, and freeways, water lines, and power lines cross it at several points.

The 1868 event was a moderate earthquake, based on the extent of the damage, probably a magnitude-6.8 earthquake. According to the Working Group, which considered, among other things, how fast stress is accumulating along the Hayward Fault, there is a 31% chance of a repeat of the 1868 earthquake or a larger event in the next 30 years. A larger event would probably involve rupture along the Rodgers Creek and possibly the Maacama Faults to the north, or rupture along the

* Here it must be pointed out that, though a specific fault might rupture only once every few thousand years, the fact that there are more than 700 active faults in California means that some "unlikely" earthquakes are expected to occur during a person's lifetime.

Green Valley-Concord and the Greenville Faults to the east, or the Calaveras Fault to the south. The group also noted that the city of San Francisco is almost equal distance from the Hayward and San Andreas Faults, so a significant earthquake along the Hayward Fault could produce shaking in San Francisco as severe as in 1906.

Elsewhere in northern California, a major earthquake along the subduction zone between Cape Mendocino and Vancouver Island—a region known to geologists and seismologists as Cascadia and which the Working Group gave a 10% chance of rupturing in the next 30 years—will almost certainly be followed within decades, perhaps even within hours, by a major earthquake along the northern segment of the San Andreas. Such an earthquake-pair sequence—rupture of the Cascadia subduction zone followed by rupture of the northern San Andreas Fault—has happened 14 times in the last 3,000 years.*

Before considering where the seismic risk is highest in the densely populated regions of California, it is important to note that there is a significant risk of one or more major earthquakes along the Walker Lane Seismic Zone in the eastern part of the state. In particular, the Working Group identified the Carson Range, Mammoth Lakes, Owens Valley, and Death Valley as places where one or more major earthquakes might occur—giving a probability of 4% in the next 30 years—potentially causing damage in the greater Reno or greater Las Vegas areas.

In southern California, the San Jacinto Fault, which runs from Cajon Pass through Riverside and continues to the southeast, was

* To emphasize a point made in a previous chapter: Contrary to popular opinion, when a major earthquake happens, the chance of another major event happening soon after does not decrease, but *increases* dramatically. For example, for any three-day period, the chance of a major earthquake occurring somewhere in California is 1 in 100,000; however, if a major earthquake has just happened, then the chance of another earthquake of *equal or greater* magnitude striking in the next three days is 1 in 10, a sobering statistic and one that needs to be known by anyone involved in rescue operations after a major seismic event.

identified by the group as a "tectonic time bomb." In fact, according to the 2008 report, the probability that this fault will rupture in the next 30 years is the same as for the Hayward Fault—31%. In either case, the result will cause extensive damage and disrupt millions of lives.

But neither the Hayward nor the San Jacinto Fault represents the greatest seismic risk in the state. That distinction, so say the experts in the Working Group—who have evaluated the geologic and paleoseismic work, examined the current levels of seismicity, have conducted extensive geodetic surveys checking to see how fast the North American and Pacific plates are moving today, and have done calculations using Coulomb-Navier stress equations—belongs to another feature, one that, according to the same experts, is "the most dangerous fault" in California.

The Palm Springs Aerial Tramway, the steepest cable ride in the United States, carries passengers up 6,000 feet from the sweltering heat of Palm Springs to the refreshingly cool and at times snow-covered hillsides close to the summit of Mount San Jacinto. The 12-minute ride gives one ample opportunity to study the mountain face and take note of the rapid change in flora from a desert floor to an alpine peak. One is also given the opportunity to look in the other direction—the entire floor of the tramcar rotates, making two revolutions during the ascent—so one is treated to an increasingly expansive view of the rugged Sonoran Desert. In the distance, from the top of the tramway one can see Mount Charleston 200 miles away near Las Vegas. To the south is the Salton Sea. Immediately in front, seemingly at one's feet, is the broad Coachella Valley, the northern extension of the Salton Trough. And running close to the axis of Coachella Valley is the San Andreas Fault.

The fault is easy to see. To the northeast, on the valley floor, is a dark green patch, the community of Desert Hot Springs. This and many other oases in Coachella Valley exist because impermeable fault gouge along the San Andreas Fault has forced groundwater to rise close to the surface.

South of Desert Hot Springs is a 20-mile-long ridge, Indio Hills. Here the San Andreas Fault consists of two strands, running on either side of the ridge. Indio Hills exists because earthquakes along both strands have pushed up the ridge.

Near the southern end of Indio Hills, the two strands merge at another oasis, Biskra Palms. From there, the San Andreas Fault is a single strand, its trace easily identified by following the eastern edge of the dark patch of irrigated fields that surround the farming communities of Indio, Coachella, Thermal, and Mecca. South of Mecca, the fault continues in a straight line to its southern end at Bombay Beach on the east side of the Salton Sea.

In all, nearly 100 miles of the San Andreas Fault can be seen from the top of the Palm Springs Aerial Tramway. It is this, the southernmost segment of the fault, that worries seismologists because this is the only segment that has not ruptured in historical time. And there are additional reasons to be concerned.

Four deep trenches have been dug across this segment, and from a detailed examination of the walls, paleoseismologists have identified five, possibly seven, major ruptures that have occurred during the last 1,100 years, the most recent in 1690.

Moreover, this is one of the most seismically active segments of the San Andreas Fault. More than a dozen moderate earthquakes have occurred since 1935. Such persistent activity could presage a major event—in the same way that a wooden board broken over a knee begins to crack before it breaks.*

To add to the concern, geodetic measurements across this segment of the fault show that points on opposites sides of the fault are

* More than a dozen moderate earthquakes occurred in the San Francisco area during the 70 years before the 1906 earthquake, but only one during the 70 years after the event.

sliding slowly and continuously—those on the west side of the fault moving to the north and those on the east side to the south—at an average rate of 1.5 inches a year. This means that, since the last major earthquake in 1690, 27 *feet* of crustal movement has accumulated on opposites sides of the fault—without any movement yet along this part of the fault. That is a buildup of an enormous amount of seismic energy that has yet to be released.

All in all, the paleoseismic evidence of five or seven major earthquakes in the last 1,100 years—showing that large events are not unusual—and the current high level of seismicity and the steady buildup of seismic energy along the fault point to one thing: A major earthquake will occur soon along the southernmost segment of the San Andreas Fault.

Thomas Jordan, director of the Southern California Earthquake Center at the University of Southern California located in Los Angeles, and a member of the group that produced the 2008 report and an update, has put it bluntly: This segment of "the San Andreas Fault is locked and loaded and ready to rumble."

But when?

Jordan and others say there is a 59% chance that a magnitude-6.7 or larger earthquake will occur along the Desert Hot Springs–Salton Sea segment of the San Andreas Fault in the next 30 years. That is almost twice the probability for a comparable earthquake happening on the nearby San Jacinto Fault or on the Hayward Fault.

Jordan and others have also considered what they have termed a "doomsday" scenario in which the entire southern half of the San Andreas Fault—from the Salton Sea to Parkfield—ruptures as a single earthquake. Parkfield is considered the northern limit because the rupture of a single event will probably not be able to propagate farther north where the stress on the fault is being relieved constantly by fault creep—evident by the slow pulling apart of the walls of the DeRose Winery—and by the frequent occurrence of magnitude-6 earthquakes, four in the last 100 years with the most recent in 2004.

Such a "wall-to-wall" rupture would involve 350 miles of the fault, considerably more than the 270 miles that ruptured in 1906, and the earthquake would be proportionally much bigger. Jordan and others estimate that such a cataclysmic event would correspond to a magnitude-8.2 earthquake and release about ten times more energy than the one in 1906.

Fortunately, the probability of such an event is low: less than 1% during the next 30 years. But considering that shaking would last more than a minute and be severe both close to the fault and in communities built over sedimentary basins—which would include but not be limited to San Bernardino, Los Angeles, and many towns in Ventura County, where, according to Jordan, the ground will shake "like a bowl of jelly"—there is still reason for concern.

Because small earthquakes are more common than large ones, a more likely scenario is that only the southernmost segment of the San Andreas Fault will rupture, at least at first, relieving some of the built-up seismic energy but not all. To understand what may follow, it is important to return to the North Anatolian and the Xiashuihe Faults and compare them to the San Andreas Fault.

All three are transform faults along plate boundaries. In all three cases, the relative plate motions on either side of the faults are the same, about 1.5 inches a year, so stress is increasing along the faults in all three places at the same rate. All three have significant strands that split off a main strand—in California, the Hayward and the San Jacinto Faults; in China, the Longmenshan; and in Turkey, north and south strands that run west of the city of Düzce and that are responsible for the creation of the sea lane known as the Dardanelles. There is also a segment of the North Anatolian Fault that creeps, just as the San Andreas Fault does north of Parkfield. Whether the Xiashuihe Fault also has a creeping segment is not known; that fault has not been studied as intensely as the other two. And two of these faults have had earthquake storms. By analogy, it seems a third earthquake storm along the San Andreas is possible.

How would such a storm evolve?

Again, from studies of the North Anatolian and the Xiashuihe Faults, it would probably begin at one end, perhaps by a rupture of the southernmost segment of the San Andreas, and proceed along the main strand and some adjacent faults.

The initial rupture would change the stress pattern—just as the 1992 Landers earthquake did and which led to the 1999 Hector Mine earthquake—so that there would be new places where stress was now concentrated. Eventually, because so much stress would be released in southern California, stress concentrations would form in northern California, jumping the 100-mile-long creeping section north of Parkfield. So the northern segment of the San Andreas Fault would also be involved.

Here a new concern arises. As the stress pattern along the San Andreas Fault changes with each successive earthquake, so does the stress pattern change along adjacent faults, causing some of them to rupture out of their previous "pattern." For example, in 1939, the earthquake that leveled the city of Erzincan broke along the main strand of the North Anatolian Fault, as well as the nearby Sungurlu-Ezinepazari fault. In China in 2008, after decades of earthquakes along the Xiashuihe Fault, a rupture occurred along a nearby parallel fault, the Longmenshan Fault. The same would happen in California.

In particular, in southern California, the Cucamonga Fault, which runs west from Cajon Pass and along the southern base of the San Gabriel Mountains, could rupture simultaneously with or soon after a major earthquake along the San Andreas Fault. And that would lead to stress changes along the Raymond Fault—which is at the western end of the Cucamonga Fault—and from that to other faults in the Los Angeles region.

In northern California, the Calaveras Fault splits off the main strand of the San Andreas just south of San Juan Bautista. So a rupture of the northern San Andreas Fault could lead to a rupture of the Calaveras—or the Hayward or the Greenville or the San Gregorio Fault.

All this is to emphasize an important point: The exact sequence of future ruptures, and hence major earthquakes, along the San Andreas and its many adjacent faults cannot be predicted—which is why Jordan and others issued probabilities in their reports. The series of quakes would not disseminate out in a necessarily coherent direction.

But one thing is certain: The last 100 years in California—which happen to correspond to a period of rapid urban growth—have been a period of seismic calm. That cannot continue.

The damaging earthquakes that have occurred—1925 Santa Barbara, 1933 Long Beach, 1952 Long Beach, 1952 south of Bakersfield, 1971 San Fernando, 1989 Loma Prieta, 1992 Landers, and 1994 Northridge—were moderate events, seismically speaking; they released only a minuscule amount of the stress that has built up between the North American and Pacific plates. This enormous amount of stress and, thus, seismic energy can only be relieved one way: as a series of large earthquakes.

And that could occur as an earthquake storm.

EPILOGUE

BODEGA BAY

What we want is a story that starts with an earthquake
and works its way up to a climax.
—Samuel Goldwyn, Hollywood movie producer

There are dozens of places where I go to sit next to the San Andreas Fault. My favorite is on the rocky headland at Bodega Bay, 60 miles north of San Francisco.

Here the fault runs close to the shoreline, making its way onto land near the marina at the north end of the bay. There is cinematographic history here: In 1963, Alfred Hitchcock filmed one of his most famous movies, *The Birds,* at the village of Bodega Bay.

Early in the film, actress Tippi Hedren is alone in a powered skiff crossing from the far side of the bay. Just before she reaches the marina, she is violently attacked by a large bird. It is the first such encounter in the movie. If one pays careful attention to the editing, one sees that the attack occurs exactly when she is over the San Andreas Fault. One can only imagine that this was a coincidence. Hitchcock, though a master of suspense and intrigue and irony, intended the village of Bodega Bay to be a seaside town in New England, where earthquakes are rare and no one ever speaks of active faults.

There is another oddity relevant to the story of earthquakes at Bodega Bay. In driving from the village to the rocky headland, one passes a deep, huge hole that was blasted into the hard granite. This hole was to be the beginning of the construction of the nation's first commercial nuclear reactor. The site is only one mile west of where the San Andreas Fault shifted in 1906. Fortunately, enough was understood about the San Andreas Fault and about earthquakes in the 1960s to halt the construction.

The white granite of the headland is comprised of large flakes of brown mica and black biotite and crossed through with dikes of pink feldspars. This granite is just one of many blocks torn out of the Earth's crust south of the Sierra Nevada Mountains and transported, in stuttering steps, hundreds of miles. These blocks now lie scattered along the California coast. Each step was a possible catastrophe, the product of an ancient earthquake.

I first visited Bodega Bay in a winter month when no one was around. So I took it upon myself to pick up the heaviest piece of loose granite that I could lift, and I carried it 15 feet to the northwest. That is about how much it will move when the 1906 earthquake is repeated. In short, I gave that boulder a head start. It has been my only real contribution to being a tectonic force.

I visit that boulder frequently. When I do, I imagine the journey it has been on—and the journey it is yet to take.

I then sit quietly and look for a long time to the southeast, wondering if a segment of the San Andreas Fault might have just ruptured. And, if it has, whether seismic waves, strong enough to hurl down a house and devastate a city, might be racing at me at unbelievable speeds.

ACKNOWLEDGMENTS

I owe thanks to many people who gave unselfishly of their time to educate me about earthquakes.

James Dieterich has been a stable influence in my life for nearly four decades and gave me my first opportunity to examine the San Andreas Fault in detail. Clarence Allen was a quiet motivator, who, when I was a student, showed me field evidence that crustal blocks had slid great distances in California. Clarence is always ready with a bag of one-liners to entertain students.

Douglas Morton can be shown a rock fragment from anywhere in southern California and tell you where the rock originated and the journey that rock has taken. I spent a pleasant day listening to him describe the endless subtleties of California geology. Ross Stein sees earthquakes in mathematical form, an ability that I envy. Ray Weldon has dug into the San Andreas Fault at several places. It was he who pointed me to the Triassic monzogranite in the San Gabriel and San Bernardino Mountains. James Dolan's enthusiasm for sleuthing ancient earthquakes is addictive. A large chunk of fault gouge and drilling mud sits on his desk.

Carol Prentice is a human encyclopedia of the 1906 earthquake and of earlier events that have occurred along the San Andreas

Fault. Tom Parsons examines patterns of earthquakes, both near California and far away. Amos Nur had an insight 20 years ago at Mycenae that led to a new understanding of when and where major earthquakes occur.

My first trip to Moonstone Beach was under the guidance of Robert Sharp. Eugene Shoemaker took me into the Mojave Desert and showed me Rainbow Basin. I can still hear his infectious laugh echo off the walls. I can also recall the high-pitched raspy voice of Jerry Eaton as he explains how he successfully challenged the building of a nuclear reactor at Bodega Bay. These three men have passed away. They are greatly missed.

Marcia McNutt and I shared an office soon after each of us completed graduate school. We talked about physics and Stoicism and whether stray cats could be used to predict earthquakes.

Also in the office was John Langbein. He and Malcolm Johnston kept the Parkfield experiment alive after the prediction window had closed.

Tom Heaton and I played basketball almost every afternoon when we were students. It was Tom who told me the story of sharing a radio interview with Charles Richter.

Special thanks are owed to Rebekah Kim and Yolanda Bustos, who opened the doors of the archives at the California Academy of Sciences. It was within those archives that I read the datebook diaries of Alice Eastwood.

Robert Tilling and David Hill have allowed me to engage them for many years in discussions about earthquakes and volcanoes and how best to communicate science to the public. Wilfred Tanigawa showed me how it is possible to glance at a seismograph and decipher the physical parameters of an earthquake.

This book has benefitted greatly from several other books. *California Earthquakes* by Carl-Henry Geschwind recounts the history of earthquake studies in the United States. *Richter's Scale* by Susan Elizabeth Hough is a passionate telling of the complex life of Charles Richter. Chapter 11 of this book is a postscript to *Assembling California* by John McPhee. *Plate Tectonics: An Insider's History of the Modern Theory of the Earth* by Naomi Oreskes tells the personal stories that led to the development of the theory of plate tectonics. And *Apocalypse* by Amos Nur shows how one person can bring a new perspective to a problem and thereby challenge established scientific thought.

Permissions to reprint photographs were provided by Susan Snyder of the Bancroft Library at the University of California, Berkeley; Christopher Crosby of the Open Topography Project at UNAVCO; John Nakata and Charles Meyer of Sight and Sound Productions in Palo Alto, California; and Andrew Selkirk, a Fellow of the Society of Antiquaries of London.

My literary agent, Laura Wood, was able to take a long-struggling author and turn him into a published one. My editor, Jessica Case, provided clarity and direction to an early manuscript that was often muddled. Both contributed immeasurably to this book.

Tom Peek has mentored my writing and has been a close friend for many years. He was living in Santa Cruz during the 1989 Loma Prieta earthquake and, at my urging, has repeated his firsthand account of what it is like to experience strong seismic shaking.

Finally, we all must acknowledge the tireless work being performed by those who are preparing California for future seismic disasters. Such disasters will certainly come.

INDEX

INDEX